# Functional Polymers and Nanomaterials for Emerging Membrane Applications

This book provides an overview of the development and selection of functional polymers and nanomaterials for membrane development and their applications. It covers the definition, classification, and preparation of various functional polymers and nanocomposites, and highlights the potential applications of functional polymers and nanomaterials in membrane technology.

Features:

- Details the selection of structural and functional materials, as well as material synthesis, modification, and characterization techniques
- Describes emerging applications of functional materials in wastewater treatment, desalination, energy, and bioremediation
- Includes numerous industrial case studies, practical examples, and questions, providing a comprehensive introduction to the topic
- Discusses industrial potential, implementation, and limitations

By combining aspects of both science and technology, this book serves as a useful resource for scientists and engineers working on membrane applications of materials.

# Emerging Materials and Technologies
*Series Editor: Boris I. Kharissov*

The *Emerging Materials and Technologies* series is devoted to highlighting publications centered on emerging advanced materials and novel technologies. Attention is paid to those newly discovered or applied materials with potential to solve pressing societal problems and improve quality of life, corresponding to environmental protection, medicine, communications, energy, transportation, advanced manufacturing, and related areas.

The series takes into account that, under present strong demands for energy, material, and cost savings, as well as heavy contamination problems and worldwide pandemic conditions, the area of emerging materials and related scalable technologies is a highly interdisciplinary field, with the need for researchers, professionals, and academics across the spectrum of engineering and technological disciplines. The main objective of this book series is to attract more attention to these materials and technologies and invite conversation among the international R&D community.

**Nanomaterials for Energy Applications**
*Edited by L. Syam Sundar, Shaik Feroz, and Faramarz Djavanroodi*

**Wastewater Treatment with the Fenton Process**
Principles and Applications
*Dominika Bury, Piotr Marcinowski, Jan Bogacki, Michal Jakubczak, and Agnieszka Jastrzebska*

**Mechanical Behavior of Advanced Materials: Modeling and Simulation**
*Edited by Jia Li and Qihong Fang*

**Shape Memory Polymer Composites**
Characterization and Modeling
*Nilesh Tiwari and Kanif M. Markad*

**Impedance Spectroscopy and its Application in Biological Detection**
*Edited by Geeta Bhatt, Manoj Bhatt and Shantanu Bhattacharya*

**Functional Polymers and Nanomaterials for Emerging Membrane Applications**
*G. Arthanareeswaran, Pei Sean Goh, and S. A. Gokula Krishnan*

For more information about this series, please visit: www.routledge.com/Emerging-Materials-and-Technologies/book-series/CRCEMT

# Functional Polymers and Nanomaterials for Emerging Membrane Applications

G. Arthanareeswaran, Pei Sean Goh,
and S. A. Gokula Krishnan

CRC Press
Taylor & Francis Group
Boca Raton London New York

CRC Press is an imprint of the
Taylor & Francis Group, an **informa** business

First edition published 2024
by CRC Press
2385 Executive Center Drive, Suite 320, Boca Raton, FL 33431

and by CRC Press
4 Park Square, Milton Park, Abingdon, Oxon, OX14 4RN

*CRC Press is an imprint of Taylor & Francis Group, LLC*

© 2024 G. Arthanareeswaran, Pei Sean Goh, and S. A. Gokula Krishnan

ISBN: 978-1-032-48908-7 (hbk)
ISBN: 978-1-032-48909-4 (pbk)
ISBN: 978-1-003-39136-4 (ebk)

DOI: 10.12019781003391364

Typeset in Times New Roman
by Apex CoVantage, LLC

# Contents

# *Preface*

Most of the commercial polymers and nanomaterials are inert and hence have limited application in various fields. Surface modifications can improve physicochemical characteristics such as adhesion, wetting, and biocompatibility. Nanomaterial and polymer functionalization enables specific moieties and bonding molecules, improving material performance. Over the last few decades, various academic and industrial researchers have been working on developing functional nanomaterials and polymers by introducing various functional groups, including hydroxyl, carboxyl, sulfonic, amine, imine, and phosphonic groups, on surfaces. The necessity for sustainable development has resulted in new objectives in nanotechnology and membrane technology, such as more effective or accurate separation and lower permissible limits for contaminant discharge into the environment. Fouling is one of the major concerns in membrane technology; to overcome fouling, life span, and membrane costs, extensive research has been conducted on the modification of membrane surface either using functional polymers or nanomaterials. Furthermore, this book offers brief knowledge on the various methods followed to synthesize functional polymers and nanomaterials, and for membrane fabrication with functionalized polymers and nanomaterials for various environmental applications like wastewater treatment, desalination, bioremediation, and fuel cell applications.

The primary audience includes typically graduate students, researchers, scientists, and membrane professionals. This advanced textbook provides an impressive overview of all aspects of selecting functional polymers and nanomaterials for membrane development and their applications. However, an authored book on the promotion of functional polymers and nanomaterials for membrane technology in wastewater treatment, bioremediation, desalination, and fuel cells has not been well elucidated yet. Several strategies have been reported for synthesizing functional polymers and materials of the membrane, including surface assembling, interfacial polymerization, electrospinning, solvothermal deposition, and layer-by-layer self-assembly. Examples are given which illustrate the state-of-the-art in the synthesis of functional polymers and nanomaterials for membranes with controlled properties. The future possibilities and limitations on synthesis, fabrications, and application are discussed. The reader is provided with references to more extended functional polymers and nanomaterials in the literature. The potential and focused areas for future innovation are indicated. By combining aspects of science and technology, this book is a useful source of information for scientists and engineers working in this field. It also provides some observations of important investigators who have contributed to the development of this subject. Integrating nanoparticles into polymers face many challenges such as immiscibility and weak interactions. Therefore, a series of novel functional polymers and nanomaterials are proposed in fabricating membranes to further enhance the desired properties. Inorganic, carbon-based membranes with tailored functionalities have more advantages of highly ordered pore structures in the development of membrane fabrication. Self-standing nanoparticles with tunable pore size, surface functionality, and defect-free structure will prove the self-standing application for water treatment, bioremediation, and fuel cell applications. Furthermore, this free-standing offers a controllable and scalable way to fabricate inorganic and carbon-based nanoparticle-integrated membranes.

This book is divided into two major sections. The first section defines, classifies, and prepares various functional polymers and nanocomposites. The selection of structural materials and functional materials will be discussed. This section also includes revolutionary material synthesis, modification, and characterization techniques. Advanced approaches toward novel processes of polymers for membranes, the synthesis of polymers with a well-defined structure, and surface functionalization will be discussed in detail in this book. The self-assembly of the structures and functions of nanomaterials will be considered in the chapters with respect to improved performance, as well as the potential implications of the developments for the future of membrane technology. The second section of this book will highlight the potential applications of functional polymers and nanomaterials in membrane technology. Here, the contribution of polymer chemistry, materials, and engineering is very important for further progress in various fields. The methodologies and approaches will be shown with additional functions, such as selective barriers with high performance, a combination of membranes with catalysis, and the modification of adaptable surfaces utilising various strategies. It is possible to create "biomimetic" membranes by combining functional macromolecular arrangements with stimuli-responsive membranes that perform an interplay of pore structure. With further advancement, these innovations made in material advancement are expected to fulfill current chemical process and membrane demand with better results. Emerging applications of functional materials in bioremediation will be highlighted in this book. Their industrial potential and implementation will also be covered.

Dr. G. Arthanareeswaran, Ph.D.
*National Institute of Technology,*
*Tiruchirappalli*

Dr. Pei Sean Goh, Ph.D.
*Universiti Teknologi Malaysia*

Mr. S. A. Gokula Krishnan, M.Tech
*National Institute of Technology,*
*Tiruchirappalli*

# *Authors*

**Professor G.Arthanareeswaran** was born in Jayankondam, India. He received the bachelor's degree in Engineering from Department of Chemical Technology, Bharathiyar University, India and received the Master degree and Doctorate in Chemical Engineering from Anna University, Chennai, India. He was a faculty at Anna University, Chennai India from 2005 to 2007. Then, he moved National Institute of Technology, Tiruchirappalli (NIT-Trichy), Tamil Nadu. India and established Membrane Research Laboratory in 2007. Now, he is professor in Department of Chemical Engineering, NIT-Trichy. He served as visiting research professor at Universiti Teknologi Malaysia, Malaysia, and Konkuk University, South Korea. To date, he has authored and co-authored more than 170 refereed journal publications, 2 books, and 7 book chapters and has been published two Indian patents. He is the principal investigator for several national and international research funding in the research area of Membrane Science and Technology. He is a fellow of the Royal Society of Chemistry, an Australian Endeavour Executive Awardee, a recipient of the Brain Pool Fellow of south Korea, Hiyoshi Environmental awardee in 2017 from Hiyoshi, Japan and Research fellow of UTM, Malaysia. His research interests in membrane materials and membrane process development for CO2 removal, fuel cell, desalination, membrane distillation and membrane sensors.

**Dr. Pei Sean Goh** is an associate professor in the School of Chemical and Energy Engineering, Faculty of Engineering, Universiti Teknologi Malaysia (UTM). She received her Ph.D. in gas engineering in 2012 at UTM. Pei Sean is also a research fellow of the Advanced Membrane Technology Research Centre (AMTEC), UTM, and the head of the Nanostructured Materials Research Group at UTM. Her research interests focus on the synthesis of a wide range of nanostructured materials and their polymer-based composites for membrane-based separation processes. Her research mainly focuses on the application of carbon-based nanomaterials and polymeric nanocomposite membranes for acidic gas removal, desalination, and wastewater treatment. Pei Sean is an associate member of the Academy of Sciences Malaysia and the winner of L'Oreal-UNESCO for Women in Science Malaysia 2020.

**Dr. S. A. Gokula Krishnan** is pursuing his Ph.D. in the Department of Chemical Engineering at the National Institute of Technology, Tiruchirappalli. He completed his bachelor's degree with a specialization in biotechnology from Anna University, India, in 2016. He completed his postgraduate degree with a specialization in industrial biotechnology from the National Institute of Technology Karnataka, Surathkal, Mangalore, India, in 2018. Later, he joined as a JRF-cum-Ph.D. in the Indo-Hungarian Joint Collaborative project in November 2018 under the guidance of Prof. Dr. G. Arthanareeswaran, Department of Chemical Engineering, National Institute of Technology Tiruchirappalli, India. Currently, he is working in the research area of photocatalytic membrane development for wastewater treatment. He has published seven research articles in highly reputed international peer-reviewed journals.

# 1

## Introduction of Functional Nanomaterials and Polymers

### 1.1 Nanotechnology

Since 2010, nanotechnology has gained huge attention in various fields like pharmaceuticals, electronics, transport, aerospace, and material manufacturing. Nanoscience and nanotechnology are, respectively, the study and use of especially small nanoscale objects with applications in chemistry, physics, materials engineering, biology, etc. There is a wide variety of nanomaterials such as nanoparticles (NPs), nanotubes, nanorods, nanoribbons, nanofibers, nanospheres, nanosheets, and quantum dots. NPs are nano-sized solid particles constructed at the molecular or atomic level to achieve new or excellent physical characteristics that cannot be achieved using traditional bulk materials. In terms of properties, NPs act like perfect units. There is a crucial range or value for any material below which its properties alter drastically. The characteristics of conventional solids are different from those of particles with a diameter of less than 100 nm. We refer to a particle as being isodimensional if all of its dimensions fall inside the nanometer range, such as a silica NP that is spherical. Definitions of NPs vary by domain and material. According to theoretical definitions, NPs are collections of millions of atoms or molecules and are frequently referred to as nanoclusters or simply clusters (Ferrando et al., 2008). These molecules or atoms can be of the same or different kinds. The surfaces of NPs can be amorphous or crystalline and can serve as droplet or gas carriers. NPs have properties between bulk materials and atomic or molecular structures. It can be applied to wastewater treatment, including photocatalysis, adsorption, desalination, electrochemical treatment, fuel cell processes, bioremediation, and industrial applications (Sahu et al., 2019; Shim et al., 2018). Recent years have seen the development of numerous NPs that have been used in ecological solutions such as water purification, persistent pollution detection, and land and water reclamation (Jeevanandam et al., 2018). The emergence of polymer composites and nanomaterials in nanotechnology has been remarkable. They are extensively used in all fields due to their enhanced electrical, magnetic, mechanical, optical, and thermal properties. Nanocomposites are solid materials with at least one phase containing particles in the range of 10–1000 nm. Nanocomposite materials have become viable alternative to microcomposites and monolithic materials. However, there are several fabrication challenges regarding the control of elemental configuration and stoichiometry in the nanocluster phase. With their unique design and properties,

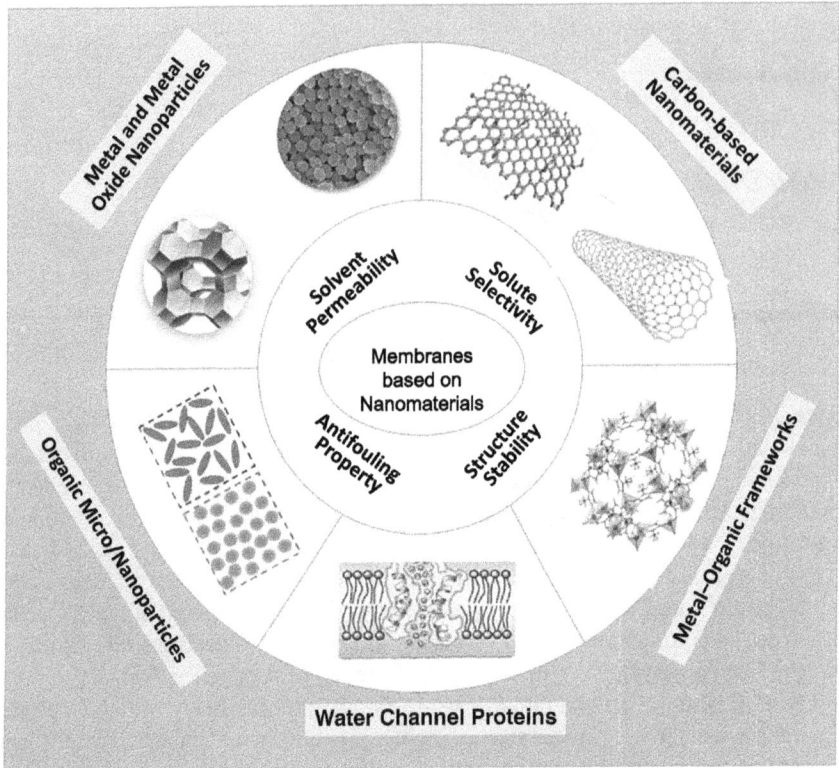

**FIGURE 1.1**
Various nanomaterials used for membrane applications (Ji et al., 2017). *Reproduced with permission from Elsevier.*

nanocomposites are appropriately noted as the material of the twenty-first century. A composite can be illustrated as an assembly of materials with various chemical and physical properties distinguished by their interfaces. Composite materials, therefore, have unique properties as opposed to distinctive materials. The majority of the composites have two phases: a continuous phase, recognized as the matrix, and a second phase, known as the dispersed phase. Nanocomposites are defined as composites with at least one of the phases in the nanoscale range (e.g., nanotubes and NPs) (Saleh et al., 2019) (Figure 1.1).

## 1.2 Polymeric Membrane Materials

Numerous contaminants can be removed from water and wastewater using membranes. The most significant drawback of modern polymer and ceramic membranes, however, is how easily they foul. Membrane fouling caused by discarded

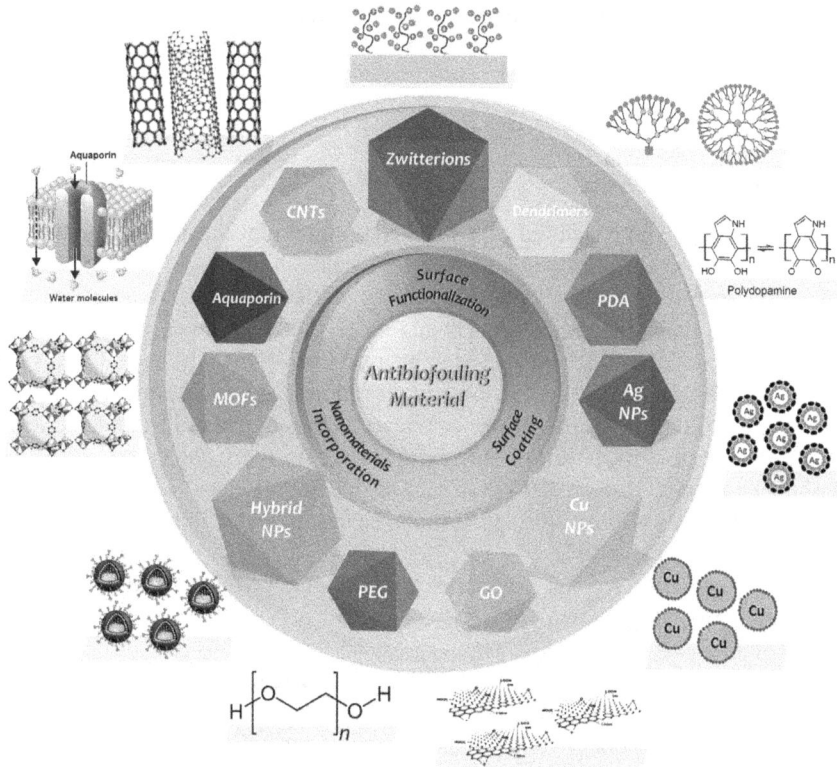

**FIGURE 1.2**
Various surface functionalization polymeric membranes (Firouzjaei et al., 2020). *Reproduced with permission from Elsevier.*

colloids, microbes, and chemicals needs to be handled carefully because it can result in increased energy usage, expensive cleaning, and membrane replacement (Committee, 2005). Intense research is being conducted in the field of membrane separation technology on the creation of low-fouling (or antifouling) membranes (Hua et al., 2008; Yang et al., 2009).

To increase the uniqueness of membrane materials, permeability and fouling resistance, and permeation quality, researchers have developed membranes utilizing CNTs or metal oxide NPs (Cortalezzi et al., 2003; J. Kim & Van der Bruggen, 2010; Li et al., 2009; Taurozzi et al., 2008). The catalytic properties of some metal oxide NPs (mainly $TiO_2$) allow the combination of chemical oxidation and NP-based membranes to mitigate membrane fouling by providing a built-in oxidation functionality. The decomposition of organic substances on the catalytic membrane's surface should also enhance permeate quality. Carbon-based nanomaterials have also been demonstrated to inactivate bacteria and viruses (Chae et al., 2009). Furthermore, it was observed that the carbon-based nanomaterials (CNTs, fullerenes, etc.) are incorporated polymeric membrane enhances the water permeability which is higher than that of pure polymer membranes (Figure 1.2).

---

## 1.3 Functional Materials

A functional material is defined as a material with certain inherent properties that allow it to perform certain functions by itself. The definition usually refers to materials with specific magnetic, photocatalytic, and electro-optical properties. Such substances are found in all materials, including organics, ceramics, and metals. It is well known that functional materials can be used for various purposes, such as fuel cell, water treatment, and desalination and biotechnology applications (I. Ali, 2012; Peng & Chen, 2009). Older and recent reviews have been material-oriented, covering cellulosic materials, phenols, GO, metal polyelectrolytes, and carbon substrates (Baibarac & Gómez-Romero, 2006). In general, the importance of advancing various functional materials for diverse applications cannot be overemphasized (Eroglu & Raston, 2017; Jia et al., 2017).

The various categories of functional materials deliberated in this chapter are listed in the following section.

(i) Semiconductor materials employed as photocatalysts are activated by absorbing photons from radiation and initiating a redox reaction. If the incident light energy is larger than or equal to the band gap of the catalytic substance, electrons in the valence band of the catalyst will be activated and move to the conduction band of the catalyst. As a result, molecules adsorbed on the surface of the photocatalyst can participate in redox processes by forming electron–hole pairs (Kumar & Chowdhury, 2020). In the advanced oxidation process, the hydroxyl ions in the system can react with the holes in the valence band to oxidize donor molecules and interact with water molecules. The hydroxyl radicals generated by this oxidation method act as strong oxidants, facilitating chemical reactions that break down contaminants in the water purification process. Conduction band electrons can undergo redox reactions and generate superoxide ions when they interact with aqueous oxygen species. The effectiveness of such semiconductor photocatalytic materials in desalination will be discussed.

(ii) Electrocatalysts assist in electrochemical reactions by interacting with an electrode's surface or, in certain cases, with the electrode itself. Through a range of electrochemical techniques, for example, electrochemical oxidation (EO), the most prominent one, electrochemistry provides a versatile and efficient way to remove different contaminants from water. Utilizing direct electron transport, electrocatalysis comprises both oxidation and reduction. Electrocatalysts play an essential role by facilitating electron transfer between the electrolyte medium and the working electrode. Thus, electrocatalysts avoid the large overvoltages required for electron delivery.

Active oxygen is formed chemically by selective oxidation of organic materials at the electrode, whereas the "active oxygen" OH radical is physically absorbed by the electrode. Therefore, these OH radicals entirely mineralize impurities and generate safe inorganic gases or ions. The suggested method for generating physically absorbed hydroxyl radicals is shown by the following chemical equation:

$$M + H_2O \rightarrow M(OH) + H^+ + e^-$$

where M stands for the anode of the electrocatalyst (Martínez-Huitle & Andrade, 2011). Electrocatalyst materials in desalination membranes will be discussed.

(iii) Magnet-based materials can be attracted or repelled when kept in a magnetic field. These functionalized materials are gaining great attention in the field of desalination because solutes are attracted during forward osmosis, selectivity is increased during reverse osmosis, and the efficiency of solar statics is enhanced (Chen et al., 2019; Razmkhah et al., 2018), to note a few.

(iv) Functional materials with antibacterial properties are important in desalination, mainly for their anti-rust characteristics. One of the main issues with membrane-based desalination processes is biofouling, which results in decreased membrane performance, increased energy use, and membrane breakdown over time. The term "biofouling" describes the accumulation and development of undesirable biological material on a membrane's surface. Research on various antimicrobial functional materials is ongoing to reduce this biofouling issue.

(v) The removal of salts and monovalent and divalent ions from water is mostly accomplished through the absorption process, which makes considerable use of absorbent materials for adsorption and absorption purposes. Adsorption is an attractive desalination technology because of its low cost, simple design, and versatility. In particular, the removal of salts and other pollutants by adsorption on selective adsorbents offers promising materials for further research in the field of adsorbents. Adsorbents, on the other hand, are crucial in solar desalination procedures because they increase the water absorption caused by solar radiation. Adsorption is a chemical's capacity to adhere to solid surfaces. During this procedure, liquid or gas molecules gather on a solid surface and interact chemically and/or physically with it to form bonds. The term "adsorbent" refers to the solid substance (Alaei Shahmirzadi et al., 2018).

In this desalination process, various functional materials are employed, each with a specific role to play in seawater and brackish water treatment. Combining multiple functions is facilitated because many materials offer multiple functions. Good examples of such materials are graphene and its composites, which can act as absorbers, photocatalysts, electrocatalysts, photothermal materials, and photocatalysts in the desalination process. This review discusses specific functional materials used in desalination and highlights existing methods, materials, and emerging trends to provide current and potential researchers with new insights about desalination (Anis et al., 2019).

## 1.4 Functional Polymers

Functional membranes are used for various purposes such as removing pollutants and dissolved particles from aqueous solutions and for adsorbing, destroying, and/ or deactivating contaminants. Typically, functional membranes are created and adapted for certain uses. Both organic and inorganic contaminants such as heavy metals, dyes, and emerging pollutants have all been eliminated using these functional membranes. It is also used in the desalination and wastewater treatment processes, as detailed in the upcoming section.

Advanced polymer processing involves fabricating functional polymer membranes. Various cutting-edge technologies have been designed to fabricate membrane matrix assemblies with a well-organized architecture and continuous membrane production. As will be discussed in the next chapters, these methods are found to be much better than those that are typically utilized in the manufacture of basic membranes. By removing undesirable connections or adding locations for new interactions, surface functionalization of membranes can also enhance material performance. Composite membranes can be created by surface modification or in situ synthesis to improve the interaction between the membrane base and additional components.

### 1.4.1 Water Treatment

In recent decades, industrial activities have greatly enhanced the amount and variety of pollutant contamination in the aquatic ecosystem, resulting in serious environmental damage (Zare, Motahari et al., 2018).

According to Bessaies et al. (2020), the majority of toxins introduced into wastewater accumulate in living things and endanger them. A range of techniques have been utilized using a variety of materials, ranging from natural to synthetic, from waste to hybrid, and from renewable energy to engineering, to remove these pollutants, which include dyes, organic contaminants, and heavy metal ions (Asif et al., 2015; Zare, Lakouraj et al., 2018). The scientific approach has evolved over the past few decades to focus on the use of bulk nanomaterials, propelling rapid advancements in nanotechnology and creating novel nanomaterials for various environmental and industrial applications. Attention to nanotechnology has been redirected due to related physicochemical properties that the bulk phase cannot contain for use in many scientific fields, particularly in water treatment (Chenab et al., 2020; Iftekhar et al., 2018). The researchers hypothesize that, over time, the presence of functional groups in polymerization, compared to bare nanomaterials, enables efficient and focused approach to address certain contaminants. Understanding the

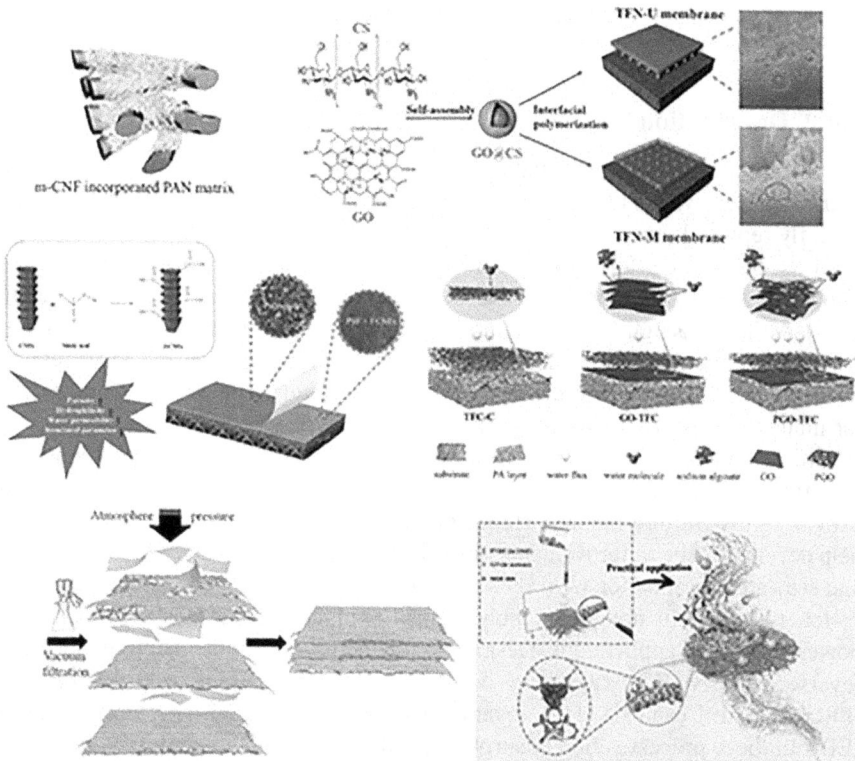

**FIGURE 1.3**
Functionalized membrane for wastewater treatment applications (A. Sahu et al., 2023). *Reproduced with permission from MDPI (Open access).*

related elimination pathways between pollutants and functional groups is necessary for selecting appropriate polymers for functionalization. Other factors, in addition to the chemical composition of functional groups, are crucial for the selectivity of polymer-functionalized materials. Physical states (such as particles, gels, fibers, and films) and physical properties should be considered for implementing a desired water treatment process (Rivas et al., 2018) (Figure 1.3). Polymeric functionalized materials for manufacturing and testing have been developed over years. N-donors (amides and amines) and O-donors (ethers and alcohols) are the commonly employed groups for functionalization and are all of significant interest. Elimination/reaction pathways depend primarily on the functional groups of the polymer material and the acidity of the effluent. For instance, anionic functional groups can remove metal cations by ion exchange and potential complexing interactions with uncharged functional groups. By changing the effluent's pH, it is easy to alter the removal method and selectivity of polymeric compounds. This section discusses how the physical states of polymeric materials and functional group characteristics affect well-water treatment approaches. The specific selectivity of functional polymeric materials with various functional groups will then be given special consideration for the target contaminants, such as carboxyl, hydroxyl, amine, phosphonic acid, and sulfonic groups (Beaugeard et al., 2020; Makvandi et al., 2021).

## 1.4.2 Desalination

Significant advances in desalination using functional materials in membrane- and thermal-based systems have been made over the past decade. For example, it was recently reported that $TiO_2$, $Al_2O_3$, $SiO_2$, ZnO, GO, reduced GO, CNTs, and MOFs used as electrolytes form the negative electrode in an electrochemical process, allowing the membrane to self-clean in situ during desalination (Hashaikeh et al., 2014). Although a significant amount of research on the numerous functional materials used in desalination processes has been published (S. Ali et al., 2019; Teow & Mohammad, 2019), each of them focuses on a particular function or on a particular material. It is crucial to do an in-depth review of the research on the various functional materials used in desalination. This study places special emphasis on how functional materials and their associated functional groups influence the field. Providing insight into the importance of using these materials in desalination will help develop further improvements toward efficient and cost-efficient brackish water and seawater desalination (Anis et al., 2019).

Ion transport in nonporous polymers is exploited in various separation and power generation applications. Desalination techniques include membrane-assisted reverse osmosis (RO), capacitive deionization (CDI), pressure-retarded osmosis (PRO), forward osmosis (FO), reverse electrodialysis (RED), and electrodialysis (ED). In these processes, ion adsorption (or distribution) is crucial. Ions must first be taken up by the polymer for ion transport through the polymer to take place, according to the solution–diffusion transport model for nonporous polymers. For the numerous applications listed earlier, sulfonated polymers have been taken into consideration. Injecting charges into the polymer backbone, such as sulfonation, can significantly affect water and salt transport properties (Geise et al., 2012). This is briefly discussed further in the following section.

## 1.4.3 Fuel Cells

Fuel cells have the advantage of providing highly efficient energy with negligible environmental pollution and can operate using unlimited sources as reactants. These advantages have allowed fuel cells to be appropriate for widespread commercial use in transportation and stationary and mobile electronics, and will therefore help meet the challenge of global energy demand (Debe, 2012). Among the currently feasible fuel cell technologies, polymer electrolyte membrane fuel cells (PEMFCs) are promising because of their low operating temperatures, high efficiency, high energy density, and fast start-up, and have been actively developed to provide power for future portable electronics, electric vehicles, and other energy-consuming devices (Rabis et al., 2012; Song et al., 2016; Zhong et al., 2008).

During PEMFC operation, hydrogen fuel is converted at the anode into protons and electrons, which travel through the external circuit to generate power.

Simultaneously, water produced by the reduction reaction of oxygen with protons is provided to the cathode through the proton exchange membrane or polymer electrolyte membrane (PEM) (Figure 1.1) (Ko et al., 2020; Wang et al., 2018). PEM has been considered the main component of PEMFC because it performs dual roles as an electrolyte and a separator, selectively transporting protons from the anode to the cathode. PEM should have great physicochemical stability and strong proton conductivity for PEMFCs to operate highly efficiently over a long period. Currently, PEMs are made of perfluorinated sulfonic acid (PFSA), and Gore-Select® (Gore), Nafion® (DuPont), Aciplex® (Asahi Chemical), Flemion® (Asahi Glass), Fumapem® (Fumatech), and Aquivion® (Solvay) have been used in commercial PEMFC systems (Souzy & Ameduri, 2005). The main advantages of PFSA polymers include (1) the hydrophobic perfluorinated backbone with a strong CF bond strength that allows for high physicochemical stability under demanding operating conditions and (2) flexible side chains that contain highly acidic fluorosulfonic acid groups that result in high proton conductivity even in low humidity conditions by facilitating the formation of ionic clusters through the interosseous phase separation structure (M. Kim et al., 2021). Despite these advantages, PEMs have a number of drawbacks, such as low glass transition temperatures that limit their operating temperatures, expensive production processes, and significant gas interference (Karimi et al., 2019).

A fuel cell, a power generation device, can directly convert the chemical energy of the fuel and oxidant into electrical energy through an electrochemical reaction. Fuel cells attract a lot of attention as one of the cleanest and most sustainable alternatives to fulfil future energy demands since they perform effectively under various operational circumstances with water as the only by-product. PEMFCs, one of several fuel cell types, are among the most suitable candidates as they are the most eco-friendly technology and feature a high energy conversion efficiency, a low operating temperature, a quick start-up time, and almost no emissions. One of the main elements of a PEMFC stack, bipolar plates constitute nearly all of the volume, 80% of the weight, and 30% of the total cost of a typical fuel cell stack. Bipolar plates have several functions. They regulate heat and water in the cells, distribute cathode and anode reactant gases uniformly, transport current between cells, and support the cell stack.

A bipolar plate should ideally have low manufacturing costs, good mechanical strength, low gas permeability, good surface qualities for moisture removal, and good corrosion resistance in the PEMFC environment. It should also have high electrical conductivity to reduce resistive losses. Bipolar plates are frequently created using graphite and graphite composites, which are both corrosion-resistant and electrically conductive. However, a number of other limiting factors, including weak reactive gas barriers, low mechanical strength, and expensive manufacturing costs, hamper their commercial implementation. When compared to materials made of graphite, metals have a significant cost, mechanical strength, ease of production, and particular volumetric power density advantage. As a result, metallic bipolar plates, particularly those made with various grades of stainless steel, have drawn a lot of interest and have the greatest potential to replace nonporous graphite, a material now employed extensively. The passive coating that forms on the surface of a

PEMFC bipolar plate when stainless steel is coated partially serves as an electrical insulator, increasing the contact resistance between the electrode and the bipolar plate. Furthermore, the weak resistance of the stainless steel to corrosion causes it to emit metal ions, including Fe, Cr, and Ni ions, which can contaminate the polymer electrolyte and Pt catalyst. Titanium and its alloys are better than stainless steel for producing metal bipolar plates for wearable applications because they can provide higher volumetric power densities when integrated into stacks. The corrosion resistance of metals can often be improved and controlled by adding protective layers to the surface without changing the bulk properties of metal substrates. In order to prevent extremely resistant oxide layers from developing on bipolar plates made of titanium and its alloys in PEMFC settings, several types of thin noble metal coatings, including gold and platinum, have been employed in recent experiments. We discovered that the performance of gold-coated bipolar plate cells was comparable to, and frequently better than, that of PEM fuel cells using graphite and uncoated titanium bipolar plates. It is important to note that in addition to providing excellent corrosion protection, protective coatings for these metal bipolar plates must also have low electrical contact resistance, high mechanical strength, excellent coating-to-substrate adhesion, and low manufacturing costs (Xu et al., 2015).

To solve these issues, a number of studies are being conducted, some of which are looking into the use of PEM based on hydrocarbons in PEMFCs. Because the chemical structure of the polymer backbone was similar to that of well-known engineering plastics, hydrocarbon-based PEM was found to have high thermal stability and mechanical strength and was also discovered to have low gas permeability due to its high crystallinity (Han et al., 2021; Harun et al., 2021). Additionally, compared to the PFSA polymer, the synthesis procedure is comparatively straightforward, and the polymer structure is easily functionalized, enabling the fabrication of tunable structures while keeping manufacturing costs low. However, sulfonated poly(ether ketone) (SPEEK), sulfonated poly(arylene ether sulfone) (SPAES), and sulfonated polyimide (SPI), which are well-known hydrocarbon polymers, are composed mainly of aromatic rings and particles. Their respective principal nucleus chains are not flexible enough to form clusters of ions by chain fractional motion. In addition, research has been conducted to improve proton conductivity by increasing the degree of protonation because the number of proton-conducting groups of these polymers (i.e., hydrocarbon-based PEMs) are relatively lower than those of PFSA-based PEMs. Hydrocarbon polymers undergo sulfonation (DS), but the water absorption ability of PEMs increases with DS, resulting in the loss of physical stability (such as dimensional stability and mechanical strength) and restriction. Due to severe deflation under actual working conditions, the use of hydrocarbon polymers in PEMFCs is constrained (Ghassemi et al., 2006). Hydrocarbon-based PEMs also have low chemical stability because the chemically unstable heteroatoms situated in the polymer backbone are prone to attack by the reactive oxygen species formed in the polymer backbone for fuel cell operation (Hwang et al., 2022; Kang et al., 2019).

Inorganic fillers and organic polymers have been used to develop new hybrid and composite membranes. This combination has enhanced a number of characteristics, including mechanical, conductive, and membrane permeability and stability, as well as affordable and ideal water retention. The internal characteristics of the particles involved, the size and type of the inorganic filler, surface alkalinity or acidity, shape, and network formation between polymer phases, as well as film fabrication techniques, all have a significant impact on the properties of hybrid membranes. Nafion is a typical membrane utilized in fuel cell applications. The potential of Nafion is undoubtedly in highly conductive materials, but it has a number of critical problems that need to be solved, such as CO poisoning, high prices, fuel crossovers, and water management. Hybrid composites are viewed as an alternative material to address the problems faced by conventional membranes and promise even greater potential if studied in detail (Zakaria et al., 2020).

Research has also been devoted to improving the long-term durability and mechanical properties of membranes by introducing inorganic materials into the ionic polymer matrix. Such investigations on pure organic polymers and inorganic nanomaterials are of interest due to the considerable changes in membrane characteristics that are seen as a result of the dual behavior and interfacial interactions of the two components. In order to increase both hydroxide conductivity and alkali stability, composite membranes can be designed by incorporating nanomaterials like $TiO_2$, $SiO_2$, CNTs, or GO into the polymer matrix (Derbali et al., 2017; Huang et al., 2017; Nonjola et al., 2013). However, few publications available are on the fabrication of composite anion exchange membranes made of fillers and polymers. Inorganic nanomaterial incorporation causes a polymer to exhibit crystalline behavior, which lowers its thermal resistance and improves fuel transfer in $H_2/O_2$ fuel cell systems. Hydrophilic inorganic additives like $SiO_2$ and $TiO_2$ have been utilized in PEMs for the purpose of self-humidification among the inorganic nanomaterials mentioned above. Due to hydrogen bonding between the functional groups of $SO_3H$ of $SiO_2$ and the polymer matrix, the composite membranes demonstrated strong electrochemical performance even at low relative humidity. Although the incorporation of fillers into polyarylene-based anion exchange membranes has been shown to improve ionic conductivity and chemical stability, the fillers are physically mixed into the polymer matrix, leading to external deformation and probably microvoid formation (Lee et al., 2021).

## References

Alaei Shahmirzadi, M. A., Hosseini, S. S., Luo, J., & Ortiz, I. (2018). Significance, evolution and recent advances in adsorption technology, materials and processes for desalination, water softening and salt removal. *Journal of Environmental Management, 215*, 324–344. https://doi.org/10.1016/j.jenvman.2018.03.040

Ali, I. (2012). New generation adsorbents for water treatment. *Chemical Reviews, 112*(10), 5073–5091. https://doi.org/10.1021/cr300133d

Ali, S., Rehman, S. A. U., Luan, H.-Y., Farid, M. U., & Huang, H. (2019). Challenges and opportunities in functional carbon nanotubes for membrane-based water treatment and desalination. *Science of the Total Environment, 646*, 1126–1139. https://doi.org/10.1016/j.scitotenv.2018.07.348

Anis, S. F., Hashaikeh, R., & Hilal, N. (2019). Functional materials in desalination: A review. *Desalination, 468*, 114077. https://doi.org/10.1016/j.desal.2019.114077

Asif, M. B., Majeed, N., Iftekhar, S., Habib, R., Fida, S., & Tabraiz, S. (2015). Chemically enhanced primary treatment of textile effluent using alum sludge and chitosan. *Desalination and Water Treatment, 53*. https://doi.org/10.1080/19443994.2015.1015448

Baibarac, M., & Gómez-Romero, P. (2006). Nanocomposites based on conducting polymers and carbon nanotubes: From fancy materials to functional applications. *Journal of Nanoscience and Nanotechnology, 6*(2), 289–302. https://doi.org/10.1166/jnn.2006.903

Beaugeard, V., Muller, J., Graillot, A., Ding, X., Robin, J.-J., & Monge, S. (2020). Acidic polymeric sorbents for the removal of metallic pollution in water: A review. *Reactive and Functional Polymers, 152*, 104599. https://doi.org/10.1016/j.reactfunctpolym.2020.104599

Bessaies, H., Iftekhar, S., Doshi, B., Kheriji, J., Ncibi, M. C., Srivastava, V., Sillanpää, M., & Hamrouni, B. (2020). Synthesis of novel adsorbent by intercalation of biopolymer in LDH for the removal of arsenic from synthetic and natural water. *Journal of Environmental Sciences, 91*, 246–261. https://doi.org/10.1016/j.jes.2020.01.028

Chae, S.-R., Wang, S., Hendren, Z. D., Wiesner, M. R., Watanabe, Y., & Gunsch, C. K. (2009). Effects of fullerene nanoparticles on Escherichia coli K12 respiratory activity in aqueous suspension and potential use for membrane biofouling control. *Journal of Membrane Science, 329*(1), 68–74. https://doi.org/10.1016/j.memsci.2008.12.023

Chen, W., Zou, C., Li, X., & Liang, H. (2019). Application of recoverable carbon nanotube nanofluids in solar desalination system: An experimental investigation. *Desalination, 451*, 92–101. https://doi.org/10.1016/j.desal.2017.09.025

Chenab, K. K., Sohrabi, B., Jafari, A., & Ramakrishna, S. (2020). Water treatment: Functional nanomaterials and applications from adsorption to photodegradation. *Materials Today Chemistry, 16*, 100262. https://doi.org/10.1016/j.mtchem.2020.100262

Committee, A. M. T. R. (2005). Committee report: Recent advances and research needs in membrane fouling. *Journal AWWA, 97*(8), 79–89. https://doi.org/10.1002/j.1551-8833.2005.tb07452.x

Cortalezzi, M. M., Rose, J., Wells, G. F., Bottero, J.-Y., Barron, A. R., & Wiesner, M. R. (2003). Ceramic membranes derived from ferroxane nanoparticles: A new route for the fabrication of iron oxide ultrafiltration membranes. *Journal of Membrane Science, 227*(1), 207–217. https://doi.org/10.1016/j.memsci.2003.08.027

Debe, M. K. (2012). Electrocatalyst approaches and challenges for automotive fuel cells. *Nature, 486*(7401), Article 7401. https://doi.org/10.1038/nature11115

Derbali, Z., Fahs, A., Chailan, J.-F., Ferrari, I. V., Di Vona, M. L., & Knauth, P. (2017). Composite anion exchange membranes with functionalized hydrophilic or hydrophobic titanium dioxide. *International Journal of Hydrogen Energy, 42*(30), 19178–19189. https://doi.org/10.1016/j.ijhydene.2017.05.208

Eroglu, E., & Raston, C. L. (2017). Nanomaterial processing strategies in functional hybrid materials for wastewater treatment using algal biomass. *Journal of Chemical Technology & Biotechnology, 92*(8), 1862–1867. https://doi.org/10.1002/jctb.5188

Ferrando, R., Jellinek, J., & Johnston, R. L. (2008). Nanoalloys: From theory to applications of alloy clusters and nanoparticles. *Chemical Reviews, 108*(3), 845–910. https://doi.org/10.1021/cr040090g

Firouzjaei, M. D., Seyedpour, S. F., Aktij, S. A., Giagnorio, M., Bazrafshan, N., Mollahosseini, A., Samadi, F., Ahmadalipour, S., Firouzjaei, F. D., Esfahani, M. R., Tiraferri, A., Elliott, M., Sangermano, M., Abdelrasoul, A., McCutcheon, J. R., Sadrzadeh, M., Esfahani, A.

R., & Rahimpour, A. (2020). Recent advances in functionalized polymer membranes for biofouling control and mitigation in forward osmosis. *Journal of Membrane Science*, *596*, 117604. https://doi.org/10.1016/j.memsci.2019.117604

Geise, G. M., Falcon, L. P., Freeman, B. D., & Paul, D. R. (2012). Sodium chloride sorption in sulfonated polymers for membrane applications. *Journal of Membrane Science*, *423–424*, 195–208. https://doi.org/10.1016/j.memsci.2012.08.014

Ghassemi, H., McGrath, J. E., & Zawodzinski, T. A. (2006). Multiblock sulfonated—fluorinated poly(arylene ether)s for a proton exchange membrane fuel cell. *Polymer*, *47*(11), 4132–4139. https://doi.org/10.1016/j.polymer.2006.02.038

Han, S.-Y., Yu, D. M., Mo, Y.-H., Ahn, S. M., Lee, J. Y., Kim, T.-H., Yoon, S. J., Hong, S., Hong, Y. T., & So, S. (2021). Ion exchange capacity controlled biphenol-based sulfonated poly(arylene ether sulfone) for polymer electrolyte membrane water electrolyzers: Comparison of random and multi-block copolymers. *Journal of Membrane Science*, *634*, 119370. https://doi.org/10.1016/j.memsci.2021.119370

Harun, N. A. M., Shaari, N., & Nik Zaiman, N. F. H. (2021). A review of alternative polymer electrolyte membrane for fuel cell application based on sulfonated poly(ether ketone). *International Journal of Energy Research*, *45*(14), 19671–19708. https://doi.org/10.1002/er.7048

Hashaikeh, R., Lalia, B. S., Kochkodan, V., & Hilal, N. (2014). A novel in situ membrane cleaning method using periodic electrolysis. *Journal of Membrane Science*, *471*, 149–154. https://doi.org/10.1016/j.memsci.2014.08.017

Hua, H., Li, N., Wu, L., Zhong, H., Wu, G., Yuan, Z., Lin, X., & Tang, L. (2008). Anti-fouling ultrafiltration membrane prepared from polysulfone-graft-methyl acrylate copolymers by UV-induced grafting method. *Journal of Environmental Sciences*, *20*(5), 565–570. https://doi.org/10.1016/S1001-0742(08)62095-1

Huang, W., Ahlfield, J. M., Zhang, X., & Kohl, P. A. (2017). Platinum supported on functionalized carbon nanotubes for oxygen reduction reaction in PEM/AEM hybrid fuel cells. *Journal of The Electrochemical Society*, *164*(4), F217. https://doi.org/10.1149/2.0191704jes

Hwang, S., Lee, H., Jeong, Y.-G., Choi, C., Hwang, I., Song, S., Nam, S. Y., Lee, J. H., & Kim, K. (2022). Polymer electrolyte membranes containing functionalized organic/inorganic composite for polymer electrolyte membrane fuel cell applications. *International Journal of Molecular Sciences*, *23*(22), Article 22. https://doi.org/10.3390/ijms232214252

Iftekhar, S., Srivastava, V., Hammouda, S. B., & Sillanpää, M. (2018). Fabrication of novel metal ion imprinted xanthan gum-layered double hydroxide nanocomposite for adsorption of rare earth elements. *Carbohydrate Polymers*, *194*, 274–284. https://doi.org/10.1016/j.carbpol.2018.04.054

Jeevanandam, J., Barhoum, A., Chan, Y. S., Dufresne, A., & Danquah, M. K. (2018). Review on nanoparticles and nanostructured materials: History, sources, toxicity and regulations. *Beilstein Journal of Nanotechnology*, *9*, 1050–1074. https://doi.org/10.3762/bjnano.9.98

Ji, Y., Qian, W., Yu, Y., An, Q., Liu, L., Zhou, Y., & Gao, C. (2017). Recent developments in nanofiltration membranes based on nanomaterials. *Chinese Journal of Chemical Engineering*, *25*(11), 1639–1652. https://doi.org/10.1016/j.cjche.2017.04.014

Jia, H., Cao, J., & Lu, Y. (2017). Design and fabrication of functional hybrid materials for catalytic applications. *Current Opinion in Green and Sustainable Chemistry*, *4*, 16–22. https://doi.org/10.1016/j.cogsc.2017.02.002

Kang, N. R., Pham, T. H., & Jannasch, P. (2019). Polyaromatic perfluorophenylsulfonic acids with high radical resistance and proton conductivity. *ACS Macro Letters*, *8*(10), 1247–1251. https://doi.org/10.1021/acsmacrolett.9b00615

Karimi, M. B., Mohammadi, F., & Hooshyari, K. (2019). Recent approaches to improve Nafion performance for fuel cell applications: A review. *International Journal of Hydrogen Energy, 44*(54), 28919–28938. https://doi.org/10.1016/j.ijhydene.2019.09.096

Kim, J., & Van der Bruggen, B. (2010). The use of nanoparticles in polymeric and ceramic membrane structures: Review of manufacturing procedures and performance improvement for water treatment. *Environmental Pollution, 158*(7), 2335–2349. https://doi.org/10.1016/j.envpol.2010.03.024

Kim, M., Ko, H., Nam, S. Y., & Kim, K. (2021). Study on control of polymeric architecture of sulfonated hydrocarbon-based polymers for high-performance polymer electrolyte membranes in fuel cell applications. *Polymers, 13*(20), Article 20. https://doi.org/10.3390/polym13203520

Ko, H., Kim, M., Nam, S. Y., & Kim, K. (2020). Research of cross-linked hydrocarbon based polymer electrolyte membranes for polymer electrolyte membrane fuel cell applications. *Membrane Journal, 30*(6), 395–408. https://doi.org/10.14579/MEMBRANE_JOURNAL.2020.30.6.395

Kumar, K., & Chowdhury, A. (2020). Use of novel nanostructured photocatalysts for the environmental sustainability of wastewater treatments. In S. Hashmi & I. A. Choudhury (Eds.), *Encyclopedia of renewable and sustainable materials* (pp. 949–964). Elsevier. https://doi.org/10.1016/B978-0-12-803581-8.11149-X

Lee, K. H., Chu, J. Y., Kim, A. R., Kim, H. G., & Yoo, D. J. (2021). Functionalized TiO2 mediated organic-inorganic composite membranes based on quaternized poly(arylene ether ketone) with enhanced ionic conductivity and alkaline stability for alkaline fuel cells. *Journal of Membrane Science, 634*, 119435. https://doi.org/10.1016/j.memsci.2021.119435

Li, J.-F., Xu, Z.-L., Yang, H., Yu, L.-Y., & Liu, M. (2009). Effect of TiO2 nanoparticles on the surface morphology and performance of microporous PES membrane. *Applied Surface Science, 255*(9), 4725–4732. https://doi.org/10.1016/j.apsusc.2008.07.139

Makvandi, P., Iftekhar, S., Pizzetti, F., Zarepour, A., Zare, E. N., Ashrafizadeh, M., Agarwal, T., Padil, V. V. T., Mohammadinejad, R., Sillanpaa, M., Maiti, T. K., Perale, G., Zarrabi, A., & Rossi, F. (2021). Functionalization of polymers and nanomaterials for water treatment, food packaging, textile and biomedical applications: A review. *Environmental Chemistry Letters, 19*(1), 583–611. https://doi.org/10.1007/s10311-020-01089-4

Martínez-Huitle, C. A., & Andrade, L. S. (2011). Electrocatalysis in wastewater treatment: Recent mechanism advances. *Química Nova, 34*, 850–858. https://doi.org/10.1590/S0100-40422011000500021

Nonjola, P. T., Mathe, M. K., & Modibedi, R. M. (2013). Chemical modification of polysulfone: Composite anionic exchange membrane with TiO2 nano-particles. *International Journal of Hydrogen Energy, 38*(12), 5115–5121. https://doi.org/10.1016/j.ijhydene.2013.02.028

Peng, B., & Chen, J. (2009). Functional materials with high-efficiency energy storage and conversion for batteries and fuel cells. *Coordination Chemistry Reviews, 253*(23), 2805–2813. https://doi.org/10.1016/j.ccr.2009.04.008

Rabis, A., Rodriguez, P., & Schmidt, T. J. (2012). Electrocatalysis for polymer electrolyte fuel cells: Recent achievements and future challenges. *ACS Catalysis, 2*(5), 864–890. https://doi.org/10.1021/cs3000864

Razmkhah, M., Moosavi, F., Mosavian, M. T. H., & Ahmadpour, A. (2018). Does electric or magnetic field affect reverse osmosis desalination? *Desalination, 432*, 55–63. https://doi.org/10.1016/j.desal.2017.12.062

Rivas, B. L., Urbano, B. F., & Sánchez, J. (2018). Water-soluble and insoluble polymers, nanoparticles, nanocomposites and hybrids with ability to remove hazardous inorganic pollutants in water. *Frontiers in Chemistry*, *6*. www.frontiersin.org/articles/10.3389/fchem.2018.00320

Sahu, A., Dosi, R., Kwiatkowski, C., Schmal, S., & Poler, J. C. (2023). Advanced polymeric nanocomposite membranes for water and wastewater treatment: A comprehensive review. *Polymers*, *15*(3), Article 3. https://doi.org/10.3390/polym15030540

Sahu, N., Rawat, S., Singh, J., Karri, R. R., Lee, S., Choi, J.-S., & Koduru, J. R. (2019). Process optimization and modeling of methylene blue adsorption using zero-valent iron nanoparticles synthesized from sweet lime pulp. *Applied Sciences*, *9*(23), Article 23. https://doi.org/10.3390/app9235112

Saleh, T. A., Parthasarathy, P., & Irfan, M. (2019). Advanced functional polymer nanocomposites and their use in water ultra-purification. *Trends in Environmental Analytical Chemistry*, *24*, e00067. https://doi.org/10.1111/j.1750-3841.2006.00195.x

Shim, J., Mazumder, P., & Kumar, M. (2018). Corn cob silica as an antibacterial support for silver nanoparticles: Efficacy on Escherichia coli and Listeria monocytogenes. *Environmental Monitoring and Assessment*, *190*(10), 583. https://doi.org/10.1007/s10661-018-6954-2

Song, Z., Cheng, N., Lushington, A., & Sun, X. (2016). Recent progress on MOF-derived nanomaterials as advanced electrocatalysts in fuel cells. *Catalysts*, *6*(8), Article 8. https://doi.org/10.3390/catal6080116

Souzy, R., & Ameduri, B. (2005). Functional fluoropolymers for fuel cell membranes. *Progress in Polymer Science*, *30*(6), 644–687. https://doi.org/10.1016/j.progpolymsci.2005.03.004

Taurozzi, J. S., Arul, H., Bosak, V. Z., Burban, A. F., Voice, T. C., Bruening, M. L., & Tarabara, V. V. (2008). Effect of filler incorporation route on the properties of polysulfone—silver nanocomposite membranes of different porosities. *Journal of Membrane Science*, *325*(1), 58–68. https://doi.org/10.1016/j.memsci.2008.07.010

Teow, Y. H., & Mohammad, A. W. (2019). New generation nanomaterials for water desalination: A review. *Desalination*, *451*, 2–17. https://doi.org/10.1016/j.desal.2017.11.041

Wang, G., Yu, Y., Liu, H., Gong, C., Wen, S., Wang, X., & Tu, Z. (2018). Progress on design and development of polymer electrolyte membrane fuel cell systems for vehicle applications: A review. *Fuel Processing Technology*, *179*, 203–228. https://doi.org/10.1016/j.fuproc.2018.06.013

Xu, J., Li, Z., Xu, S., Munroe, P., & Xie, Z.-H. (2015). A nanocrystalline zirconium carbide coating as a functional corrosion-resistant barrier for polymer electrolyte membrane fuel cell application. *Journal of Power Sources*, *297*, 359–369. https://doi.org/10.1016/j.jpowsour.2015.08.024

Yang, S., Yang, F., Fu, Z., & Lei, R. (2009). Comparison between a moving bed membrane bioreactor and a conventional membrane bioreactor on organic carbon and nitrogen removal. *Bioresource Technology*, *100*(8), 2369–2374. https://doi.org/10.1016/j.biortech.2008.11.022

Zakaria, Z., Shaari, N., Kamarudin, S. K., Bahru, R., & Musa, M. T. (2020). A review of progressive advanced polymer nanohybrid membrane in fuel cell application. *International Journal of Energy Research*, *44*(11), 8255–8295. https://doi.org/10.1002/er.5516

Zare, E. N., Lakouraj, M. M., & Kasirian, N. (2018). Development of effective nano-biosorbent based on poly m-phenylenediamine grafted dextrin for removal of Pb (II) and methylene blue from water. *Carbohydrate Polymers*, *201*, 539–548. https://doi.org/10.1016/j.carbpol.2018.08.091

Zare, E. N., Motahari, A., & Sillanpää, M. (2018). Nanoadsorbents based on conducting polymer nanocomposites with main focus on polyaniline and its derivatives for removal of
heavy metal ions/dyes: A review. *Environmental Research, 162,* 173–195. https://doi.
org/10.1016/j.envres.2017.12.025

Zhong, C.-J., Luo, J., Njoki, P. N., Mott, D., Wanjala, B., Loukrakpam, R., Lim, S., Wang,
L., Fang, B., & Xu, Z. (2008). Fuel cell technology: Nano-engineered multimetallic
catalysts. *Energy & Environmental Science, 1*(4), 454–466. https://doi.org/10.1039/
B810734N

# 2

## Classification of Functional Nanomaterials and Polymers

### 2.1 Definition

Nanomaterials (NMs) are particles that are 1–100 nm in size. NMs can have their surfaces functionalized in one of two ways: by covalent modification through a regular organic synthesis process or through noncovalent modification through complexation, adsorption, or grafting. Functionalization of nanoparticles (NPs) refers to altering or modifying the surface structure by the addition of various organic polymers onto the NP surface.

Materials that are created by the assembly or repetition of nanometer-sized building components are known as nanostructured materials. The constituent parts are referred to as grains or crystallites. Building blocks can be of zero, one, or two dimensions and can be divided by many kinds of barriers. These materials are divided into the following categories based on the crystallite shapes:

1. Materials with nanostructures and equiaxed crystallites
2. Materials with nanostructures and rod-shaped crystallites
3. Stacked crystallites in nanostructured materials
4. Compared to bulk structured materials, nanostructured materials have distinctive features.

Functionalization is the surface modification of NPs, which involves attaching substances like folic acid, biotin molecules, oligonucleotides, peptides, and antibodies to the surface of NPs to improve their characteristics and attack the target with great accuracy. Functionalized NPs also exhibit excellent physical qualities, such as anticorrosion, anti-agglomeration, and noninvasiveness. To improve the NPs' overall effectiveness and modality, extensive research has been done to functionalize them.

Functionalization improves the qualities and traits of NPs, allowing them to have a significant impact in the field of biomedical, membrane separation, fuel cell, and other applications. In molecular imaging, good functional pictures with appropriate differentiation that can be viewed with high contrast are essential in order to gather useful information. Thiruppathi et al., (2017) addresses how functionalization improves molecular imaging and enables multimodal imaging, which allows for the creation of pictures that combine the functionalities unique to each modality. This

DOI: 10.12019781003391364-2

also explains how NPs that interact with molecules at the molecular level may target certain cells or substances with great specificity, decreasing background signal and enabling concurrent therapy and imaging.

---

## 2.2  Functional Polymers

Functional polymers are macromolecules with specific features and applications. The chemical functional groups that differ from those of the backbone chains frequently control the characteristics of such materials. Examples include hydrophobic groups on polar polymer chains or polar or ionic functional groups on hydrocarbon backbones. The chemical heterogeneity of the polymer chains can cause phase separation, association, or increased reactivity. Another motivator is the capacity of functional polymers to self-assemble or build supramolecular structures. These so-called smart materials can develop when the creation or dissociation of the self-assemblies is activated by chemical or physical stimuli. The majority of functional polymers have straightforward linear backbones. They can be block, graft, chain-end (telechelic), or in-chain structures (Schulz & Patil, 1998). Various polymeric materials, including polyethersulfone (PES), polysulfone (PSF), polyvinylidene fluoride (PVDF), polyacrylonitrile (PAN), polycarbonate (PC), polyethylene (PE), polypropylene (PP), polyimide (PI), polyetherimide (PEID), polytetrafluoroethylene (PTFE), polyamide (PA), cellulose acetate (CA), and polypyrrole (PPy), are used as base polymers in the functionalization approaches.

Both in academics and business, functional polymers are becoming increasingly popular. These macromolecules have particular properties and uses. The properties of this class of materials mostly depend on the presence of chemicals such as polar or ionic functional groups on the hydrocarbon backbones or hydrophobic groups on polar polymer chains, which have different functional groups from the backbone chains. The mechanical strength, flexibility, chemical stability, and processability of polymer backbones are taken into consideration while choosing them. Chemical heterogeneity that is produced as a result of functionalizing the bulk polymer has several benefits, including increased reactivity, phase separation, compatibility, and association. Another advantage is the capacity of functional polymers to assemble themselves or form supramolecular structures (Hosseini et al., 2019).

Modern functional polymers have unique properties and rules. First, the processes used to synthesize monomers are typically intricate and varied, including organic chemistry and metal catalysis. Second, the inclusion of functional monomers necessitates the development of novel polymerization processes, techniques, mechanisms, and catalysts. Finally, their functions are strongly linked to both their aggregate states and the chemical composition of the polymer chains. Therefore, it is required to integrate their synthesis, processing, and morphology to produce polymers with superior functionalities (Figure 2.1). The multidisciplinary nature of functional polymeric materials allows for the creation of novel materials and the enhancement of the performance of already existing polymeric materials (Wang et al., 2020).

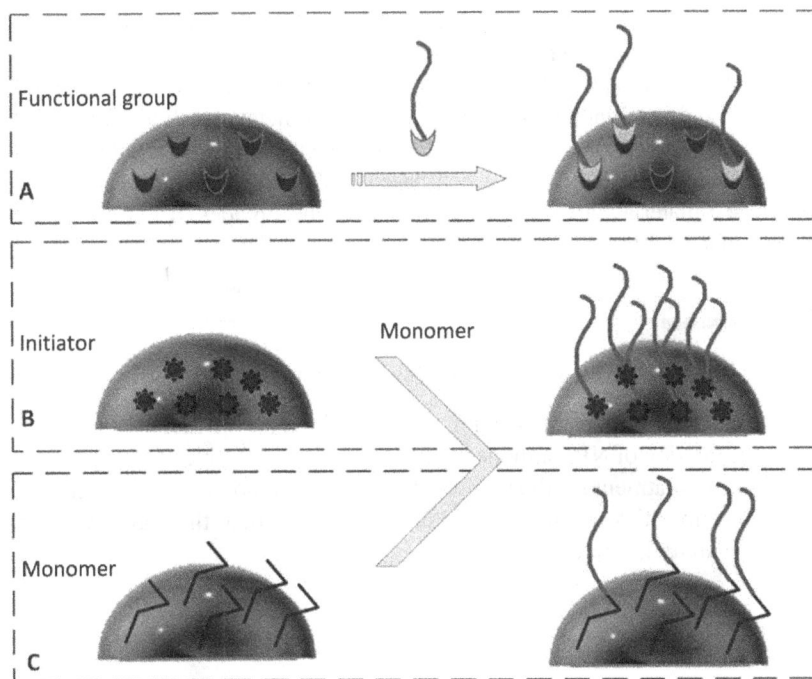

**FIGURE 2.1**
Functionalization of NPs (Wieszczycka et al., 2021). *Reproduced with permission from Elsevier.*

The attachment of organic moieties or polymers to the surface of a NP is known as functionalization. The surface of NPs is modified by the physical or chemical attachment of various organic compounds and functional moieties, enhancing the properties of the NPs. Functionalized NMs are substances with these enhanced qualities and characteristics. Due to the numerous uses for functionalized NMs in industries including for biomedicine, bioenergy, and biosensor development, they have drawn great attention (Mahajan et al., 2022).

## 2.3 Functionalization of Nanomaterials

Due to their unique characteristic features and extraordinary changes compared to bare NPs, functionalized NMs have recently attracted interest and are being applied in various industries (Hussain et al., 2018). Lack of surface reactivity and mobilization and increased aggregation are only a few of the numerous disadvantages of bare NPs. Functionalization, which aids in creating environmentally benign engineered NMs with a narrow bandgap, improved surface reactivity, stability, and mobility, can eliminate these restrictions. Advanced functionalized NMs have traits including higher chemical and antibacterial activity, and they are known for their enhanced capacity to remove toxins from water and the environment via absorption. Bromine. Functionalized CNTs are created by using the vapor phase bromination process, which involves treating multi-walled nanotubes with bromine vapors. Carbon-based NMs, if utilized combined with metals or oxides of metals (such as zinc oxide and graphene), are known to remove hazardous pollutants from wastewater. Due to the synergistic interaction between carbon-based NMs and metallic NPs, when these carbon-based materials are doped with metals or their oxides, beneficial modifications take place (Mittal et al., 2015).

## 2.4  Classification of Functionalized Nanoparticles

NPs are the most important factor for the modification of membrane structures using incorporation of NPs onto the membrane surface, which creates a pathway for efficient water treatment for the removal of organic and inorganic contaminants and varieties of impurities in water. Based on the classification, there are five types of NPs used in water treatment:

1. Carbon-based NMs
2. Organic-based NMs
3. Inorganic-based NMs
4. Metal organic framework
5. Covalent organic framework

The sub-categories of different types of NPs are discussed as follows.

### 2.4.1  Carbon-based Nanomaterials

1. Graphene oxide (GO)
2. Carbon nanotubes (CNTs)
3. Fullerenes
4. Activated carbon (AC)
5. C60

### 2.4.2  Inorganic-Based Nanomaterials

1. Titanium dioxide ($TiO_2$)
2. Bismuth tungstate ($Bi_2WO_6$)
3. Cerium dioxide ($CeO_2$)
4. Aluminum oxide ($Al_2O_3$)
5. MXene
6. Silicon dioxide ($SiO_2$)

### 2.4.3 Metal–Organic Framework

There are many types of metal–organic framework (MOF) NMs that are considered emerging materials for water treatment, which are as follows:

| | | | |
|---|---|---|---|
| 1. | MOF-5 | 6. | UiO-66 |
| 2. | MOF-88 | 7. | Nu-125 |
| 3. | MOF-177 | 8. | IRMOF-74-XI |
| 4. | Cr-MOF-101 | 9. | Ni-CPO-27 |
| 5. | Cr-MOF-100 | 10. | ZIF-8 |

## 2.5 Why Should Nanoparticles Be Functionalized?

Functionalization refers to the surface modification of NMs, which involves the adherence of various chemicals to the surface of NMs which enhances their characteristics. Functionalized NPs also exhibit excellent physical qualities, such as anti-corrosion, anti-agglomeration, and noninvasiveness. To improve the NPs' overall effectiveness and modality, extensive research is being done to functionalize them (Wang et al., 2020).

## 2.6 How Are Nanoparticles Functionalized?

NPs are functionalized using various materials by employing processes like hydrothermal/solvothermal methods, in situ polymerization, etc. By functionalization, the properties and structures of NPs are modified by several factors. In this study, we briefly discuss how NPs are functionalized and their properties and applications.

### 2.6.1 Inorganic Nanomaterials

NPs that are not formed of carbon or biological elements belong to this class. The most common examples of this class include semiconductor NPs, metal NPs, and ceramic NPs. Purely composed of metal precursors, metal nanoparticles (MNPs) can be monometallic, bimetallic, or polymetallic. Bimetallic NPs can be generated in layers (core–shell) or from alloys. These NPs have special optical and electrical capabilities because of the peculiarities of the localized surface plasmon resonance. Certain metal NPs also have special biological, magnetic, and thermal properties. These NPs also have increased importance as building blocks for nanodevices with a wide range of physical, chemical, biological, biomedical, and pharmacological uses (these applications are discussed in detail in the upcoming chapters). Modern-day cutting-edge materials require the regulated synthesis of metal NPs in terms of size, shape, and face.

## 2.6.1.1 Functionalization of Titanium Dioxide

TiO$_2$ (titanium dioxide) photocatalysis, one of the ultraviolet (UV)-based advanced oxidation nanotechnologies (AONs) and technologies (AOTs), has received more attention for the development of efficient water purification and water treatment systems due to the strength of TiO$_2$ in generating highly oxidizing hydroxyl radicals, which easily attack and disintegrate organic contaminants in water (Choi et al., 2014). TiO$_2$ is the most-used photocatalyst in water treatment owing to its lower toxicity and high chemical stability (Maiti et al., 2019). Functionalization of TiO$_2$ is done by using oleic acid, polyacrylic acid (PAA), propionic acid (PPA), polylactic acid (PLA), monoethanolamine (MEOA), triethylenetetramine (TETA), and maleic anhydride (MA). We briefly discuss the aforementioned functionalization process as follows.

### 2.6.1.1.1 PAA-TiO$_2$-functionalized Nanoparticles

The PVDF membrane's antifouling characteristics were enhanced by the grafting of PAA. TiO$_2$ NP immobilization improved this characteristic. Due to the greatest immobilization of TiO$_2$, the "grafting from" procedure produced the best fouling resistance in whey filtering (Figure 2.2). Hydrophobic adsorption of whey protein to the modified membrane surface was reduced by the hydrophilic functional groups of PAA and the photocatalytic property of TiO$_2$ (Madaeni et al., 2011).

### SPEEK/amine-functionalized

Grafting PEI with abundant amine groups onto the titania fillers remarkably increased the content of facilitated transport sites in the membranes, leading to an

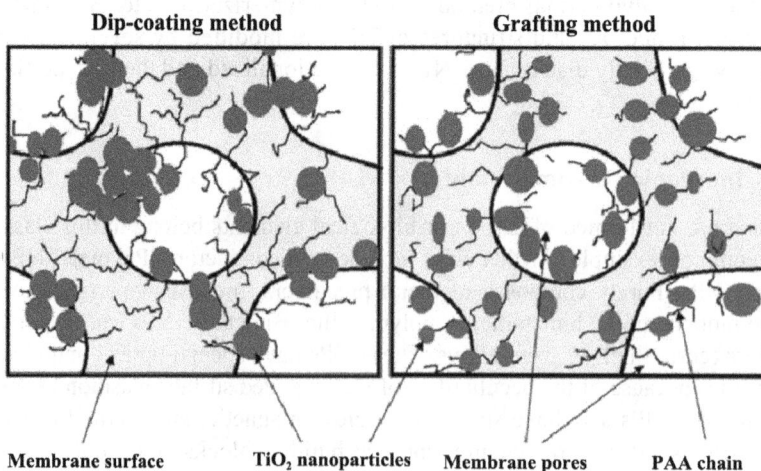

**FIGURE 2.2**
*In situ* polymerization of TiO$_2$ NPs on PVDF membrane (Madaeni et al., 2011). *Reproduced with permission from Elsevier.*

increment in both gas permeability and selectivity. High humidity also contributed to the facilitated transport of $CO_2$ via the generation of $HCO_3^-$ (Xin et al., 2014).

### 2.6.1.2 Functionalization of $SiO_2$

#### 2.6.1.2.1 Functionalized PSf/$SiO_2$ Nanocomposite Membrane

While making ultrafiltration membranes, PSF is frequently used as the material of choice. Its hydrophobic properties, however, inevitably result in subpar membrane function. To increase the permeability and antifouling capabilities of the PSF membrane, $SiO_2$ NPs were functionalized utilizing a phase inversion technique. Due to the interfacial tensions between the polymer and filler, the $SiO_2$-filled membrane's pores grew larger and linked ($SiO_2$ NPs). The addition of $SiO_2$ NPs significantly increased the modified membrane's permeability while only partially making up for its decreased NTU. The antifouling capabilities of the manufactured membrane were enhanced by the inclusion of $SiO_2$ NPs to the blended composition (Ahmad et al., 2011).

#### 2.6.1.2.2 Functionalized PVA/$SiO_2$ Nanoparticles

The "one-step" hydrolysis and co-condensation method was used to create the polyvinyl alcohol (PVA)/$SiO_2$ hybrid coatings, which had distinct mesoporous structures and a lot of amino groups. Also, the membranes as-prepared are a good candidate for use in actual wastewater treatment, bridging the gap between pollutant adsorption and oil/water separation, thanks to their exceptional reusability and significant chemical durability. The PVDF membranes have remarkable hydrophilicity, underwater superoleophobicity, and ultralow oil adhesion, thanks to the hybrid coating. The pretreated commercial PVDF microfiltration membranes were coated with mesostructured amino-functionalized PVA/$SiO_2$ hybrid gel via a straightforward dip-coating method to create microfiltration membranes with noteworthy performance in highly efficient oil-in-water emulsion separation and water pollutant adsorption (Figure 2.3). The PVA/$SiO_2$-coated PVDF membranes are attractive materials in real-world applications for sophisticated wastewater treatment since they also have great corrosion resistance and strong recyclability (Liu et al., 2019).

### 2.6.1.3 Functionalization of CNTs

Raw CNTs have frequently been subjected to oxidation with $HNO_3$, $H_2SO_4$, HCl, $H_2O_2$, $KMnO_4$, and NaOCl, or occasionally a mixture of some of these, in order to introduce oxidized functional groups. In general, oxidation reduces the dispersibility and increases the capacity to absorb some toxic pollutants in water and wastewater at the cost of slight surface damage to CNTs, as previously mentioned. Furthermore, oxygen plasma treatment can be used to modify the surface of CNTs by including functional groups that contain oxygen. Metal oxides like $Al_2O_3$, $MnO_2$, and $Fe_3O_4$ can also be used to coat the surface of CNTs, providing another method to modify the surface in addition to oxidation and plasma treatment. This increases the contaminant removal efficiency of CNTs (Aslam et al., 2021; Liang et al., 2015; Moghaddam & Pakizeh, 2015; Rao et al., 2007; Ren et al., 2011; Tang et al., 2012).

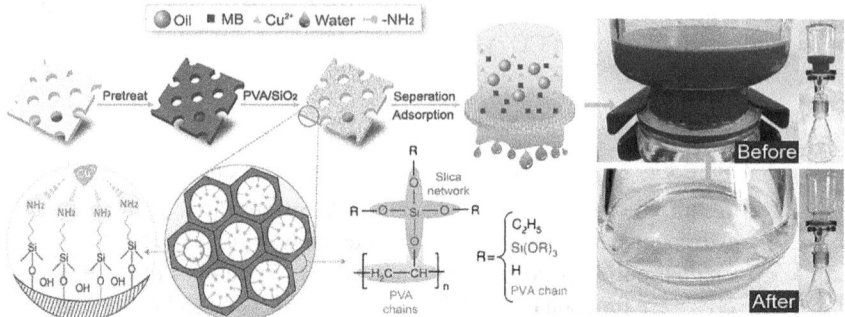

**FIGURE 2.3**
Surface functionalization of PVDF membrane with PVA/SiO$_2$ NPs (Liu et al., 2019). *Reproduced with permission from Elsevier.*

### 2.6.1.3.1 Multiwalled CNT Functionalized with Diethylenetriaminepentaacetic Acid (DTPA)

A long-term membrane operation would benefit from the modified DT-4 membrane's high water flux, antifouling characteristics, and effective separation performance. Due to its strong underwater antifouling performance, the DT-4 membrane can be reused and demonstrated long-term operation stability. The improved membrane shows excellent potential for practical applications by enabling significant oil recovery from water emulsions (Venkatesh et al., 2021).

### 2.6.1.3.2 Tannic Acid–Fe$^{III}$-functionalized Multiwalled Carbon Nanotubes (TA-MWNTs)

The surface of MWNTs was functionalized by the quick and reliable development of a stable tannic acid–Fe$^{III}$ complex coating. Due to the many hydrophilic groups in TA, functionalized MWNTs (TA-MWNTs) display good dispersity in an aqueous solution. Interfacial polymerization was then used to create the PA thin-film nanocomposite (TFN) membrane while TA-MWNTs were present. The TFN membrane has improved antifouling performance and increased long-term stability. It is important to note that the TFN membrane also has excellent chemical stability and oxidation resistance due to the ability of the phenol groups in TA to capture free radicals and the interaction between the phenol groups of TA and the acyl chloride groups of trimesoyl chloride (TMC) which results in a more stable polyester bonding (Wu et al., 2017).

### 2.6.1.3.3 Oleic Acid-functionalized CNTs

By covering a PTFE support membrane with PVDF solution containing CNTs that have been functionalized with oleic acid, a unique composite membrane was created for the vacuum membrane distillation (VMD) process. The PTFE support membrane's hydrophobic holes are crucial for allowing water vapor to pass through. According to the FTIR measurement, oleic acid modification of CNTs can lessen their hydrophilicity when compared to unoxidized CNTs. The membrane's performance in the VMD process is enhanced by the PVDF polymer that is present on the functionalized CNTs' surfaces and in the spaces between the CNTs. On the other hand, the coated active layers of CNTs become interconnected due to the presence of the PVDF polymer (Pouya et al., 2021).

### 2.6.1.4 Functionalization of Graphene Oxide

Functionalized graphene sheets with organic functional groups have been developed for a variety of uses. The primary goal is to disperse graphene in typical organic solvents, which is often accomplished by attaching specific chemical groups. A critical step in creating graphene-based nanocomposite materials is the dispersion of graphene sheets in organic solvents. Moreover, novel features offered by organic functional groups like chromophores could be integrated with graphene's conductivity-related attributes. The extended aromatic property of graphene is typically disrupted when organic molecules are covalently linked to its surface, allowing the modification of its electrical characteristics. A potent way to employ graphene is to chemically dope it in order to create a band gap in nanoelectronic devices (Georgakilas et al., 2012).

#### 2.6.1.4.1 Sulfonic Acid-functionalized GO

The increased surface abrasion, porosity, and pore size of the membranes were brought on by the addition of GO and sulfonated graphene oxide (SGO). Additional sulfonic groups ($-SO_3H$) hold a larger water hydrogen layer and enhance the water flux on SGO supports. Compared to the -COOH/-OH groups found in GO, the anchoring $-SO_3H$ group in SGO forms a stronger hydrogen bond. This was determined to be due to the strong electrostatic and hydrogen bonding forces of SGO against fouling proteins. Overall, the induced SGO NP offers a fresh way to improve PVDF ultrafiltration membranes' hydrophilicity, water flow, antifouling, and mechanical performance. The hydrophilic $-SO_3H$ groups of the membrane surface, which have a larger surface area and can adsorb a denser and more stable water layer than the COOH group of GO, were said to be responsible for this (Figure 2.4). Comparative investigations revealed that the P-SGO blended membranes outperformed the PVDF-GO and PVDF-SGO blended membranes in terms of hydrophilicity, pure water penetration flux, and antifouling capabilities due to their higher pore density, rougher surface, and lower contact angle. It is anticipated that SGO additive materials will create new possibilities for enhancing the functionality of modified membranes in ultrafiltration applications (Ayyaru & Ahn, 2017).

#### 2.6.1.4.2 Tannic Acid-functionalized GO

GO is offered as a novel hydrophilic modifier of thin-film nanocomposites (TFN) membranes after being functionalized with natural tannic acid (TA) macromolecules (GO-TA). GO-TA was added to the active layers of TFN membranes via interfacial polymerization of m-phenylenediamine and trimesoyl chloride monomers. Next, we will look at how surface modification affects the active layer's surface hydrophilicity, roughness, morphologies, and surface chemistry. The latest thin-film nanocomposites-forward osmosis (TFN-FO) membrane's potential for commercial use is increased by the good separation performance. This demonstrates the great stability and effectiveness of the GO-TA-modified rejection layer. Overall,

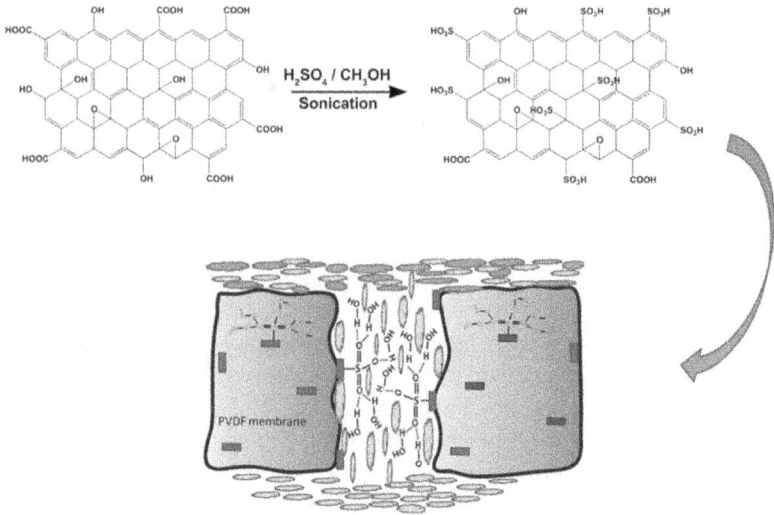

**FIGURE 2.4**
Surface modification of PVDF membrane using SGO (Ayyaru & Ahn, 2017). *Reproduced with permission from Elsevier.*

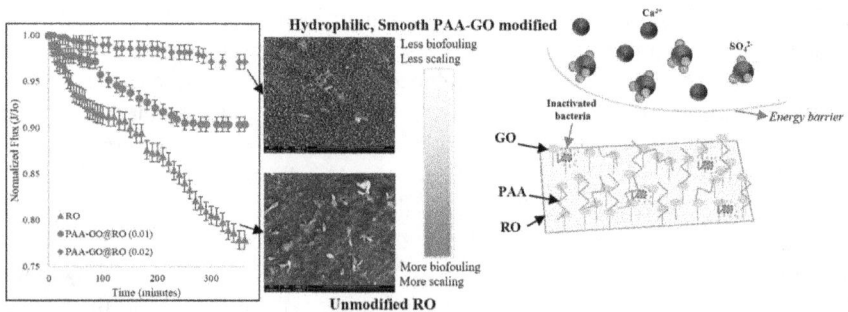

**FIGURE 2.5**
Surface functionalization of RO membrane with PAA and GO NPs (Ashfaq et al., 2020). *Reproduced with permission from Elsevier.*

the PA thin-film modification with GO-TA demonstrated that this compound has the potential to be a useful nano modifier for producing high-performance TFN-FO membranes (Yassari et al., 2022).

### 2.6.1.4.3 PAA-functionalized GO

For reducing both biofouling and mineral scaling, GO, followed by polymerization of acrylic acid (used as an antiscalant), is utilized. As a result, it was discovered that polymer-modified GO-coated RO membranes could reduce the formation of both gypsum scaling and biofilms, exhibiting their capability to manage various types of membrane fouling (Figure 2.5). The polymer-modified GO-coated RO membranes' potential to manage membrane scaling and biofouling in seawater reverse osmosis (SWRO) systems was further supported by additional research on the prevention of both gypsum precipitation and biofilm formation (Ashfaq et al., 2020).

## References

Ahmad, A. L., Majid, M. A., & Ooi, B. S. (2011). Functionalized PSf/SiO2 nanocomposite membrane for oil-in-water emulsion separation. *Desalination, 268*(1–3), 266–269. https://doi.org/10.1016/j.desal.2010.10.017

Ashfaq, M. Y., Al-Ghouti, M. A., & Zouari, N. (2020). Functionalization of reverse osmosis membrane with graphene oxide and polyacrylic acid to control biofouling and mineral scaling. *Science of the Total Environment, 736*, 139500. https://doi.org/10.1016/j.scitotenv.2020.139500

Aslam, M. M.-A., Kuo, H.-W., Den, W., Usman, M., Sultan, M., & Ashraf, H. (2021). Functionalized carbon nanotubes (CNTs) for water and wastewater treatment: Preparation to application. *Sustainability, 13*(10), Article 10. https://doi.org/10.3390/su13105717

Ayyaru, S., & Ahn, Y.-H. (2017). Application of sulfonic acid group functionalized graphene oxide to improve hydrophilicity, permeability, and antifouling of PVDF nanocomposite ultrafiltration membranes. *Journal of Membrane Science, 525*, 210–219. https://doi.org/10.1016/j.memsci.2016.10.048

Choi, H., Zakersalehi, A., Al-Abed, S. R., Han, C., & Dionysiou, D. D. (2014). Chapter 8—Nanostructured titanium oxide film- and membrane-based photocatalysis for water treatment. In A. Street, R. Sustich, J. Duncan, & N. Savage (Eds.), *Nanotechnology applications for clean water* (2nd ed., pp. 123–132). William Andrew Publishing. https://doi.org/10.1016/B978-1-4557-3116-9.00008-1

Georgakilas, V., Otyepka, M., Bourlinos, A. B., Chandra, V., Kim, N., Kemp, K. C., Hobza, P., Zboril, R., & Kim, K. S. (2012). Functionalization of graphene: Covalent and non-covalent approaches, derivatives and applications. *Chemical Reviews, 112*(11), 6156–6214. https://doi.org/10.1021/cr3000412

Hosseini, M.-S., Amjadi, I., Mohajeri, M., Zubair Iqbal, M., Wu, A., & Mozafari, M. (2019). Chapter 1—Functional polymers: An introduction in the context of biomedical engineering. In M. Mozafari & N. P. Singh Chauhan (Eds.), *Advanced functional polymers for biomedical applications* (pp. 1–20). Elsevier. https://doi.org/10.1016/B978-0-12-816349-8.00001-1

Hussain, I., Tran, H. P., Jaksik, J., Moore, J., Islam, N., & Uddin, M. J. (2018). Functional materials, device architecture, and flexibility of perovskite solar cell. *Emergent Materials, 1*(3), 133–154. https://doi.org/10.1007/s42247-018-0013-1

Liang, J., Li, L., Chen, D., Hajagos, T., Ren, Z., Chou, S.-Y., Hu, W., & Pei, Q. (2015). Intrinsically stretchable and transparent thin-film transistors based on printable silver nanowires, carbon nanotubes and an elastomeric dielectric. *Nature Communications, 6*(1), Article 1. https://doi.org/10.1038/ncomms8647

Liu, H., Yu, H., Yuan, X., Ding, W., Li, Y., & Wang, J. (2019). Amino-functionalized mesoporous PVA/SiO2 hybrids coated membrane for simultaneous removal of oils and water-soluble contaminants from emulsion. *Chemical Engineering Journal, 374*, 1394–1402. https://doi.org/10.1016/j.cej.2019.05.161

Madaeni, S. S., Zinadini, S., & Vatanpour, V. (2011). A new approach to improve antifouling property of PVDF membrane using in situ polymerization of PAA functionalized TiO2 nanoparticles. *Journal of Membrane Science, 380*(1–2), 155–162. https://doi.org/10.1016/j.memsci.2011.07.006

Mahajan, G., Kaur, M., & Gupta, R. (2022). Chapter 2—Green functionalized nanomaterials: Fundamentals and future opportunities. In U. Shanker, C. M. Hussain, & M. Rani (Eds.), *Green functionalized nanomaterials for environmental applications* (pp. 21–41). Elsevier. https://doi.org/10.1016/B978-0-12-823137-1.00003-8

Maiti, A., Mishra, S., & Chaudhary, M. (2019). Chapter 25—Nanoscale materials for arsenic removal from water. In S. Thomas, D. Pasquini, S.-Y. Leu, & D. A. Gopakumar (Eds.), *Nanoscale materials in water purification* (pp. 707–733). Elsevier. https://doi.org/10.1016/B978-0-12-813926-4.00032-X

Mittal, G., Dhand, V., Rhee, K. Y., Park, S.-J., & Lee, W. R. (2015). A review on carbon nanotubes and graphene as fillers in reinforced polymer nanocomposites. *Journal of Industrial and Engineering Chemistry, 21*, 11–25. https://doi.org/10.1016/j.jiec.2014.03.022

Moghaddam, H. K., & Pakizeh, M. (2015). Experimental study on mercury ions removal from aqueous solution by MnO$_2$/CNTs nanocomposite adsorbent. *Journal of Industrial and Engineering Chemistry, 21*, 221–229. https://doi.org/10.1016/j.jiec.2014.02.028

Pouya, Z. A., Tofighy, M. A., & Mohammadi, T. (2021). Synthesis and characterization of polytetrafluoroethylene/oleic acid-functionalized carbon nanotubes composite membrane for desalination by vacuum membrane distillation. *Desalination, 503*, 114931. https://doi.org/10.1016/j.desal.2021.114931

Rao, G. P., Lu, C., & Su, F. (2007). Sorption of divalent metal ions from aqueous solution by carbon nanotubes: A review. *Separation and Purification Technology, 58*(1), 224–231. https://doi.org/10.1016/j.seppur.2006.12.006

Ren, X., Chen, C., Nagatsu, M., & Wang, X. (2011). Carbon nanotubes as adsorbents in environmental pollution management: A review. *Chemical Engineering Journal, 170*(2), 395–410. https://doi.org/10.1016/j.cej.2010.08.045

Schulz, D. N., & Patil, A. O. (1998). Functional polymers: An overview. In *Functional polymers* (Vol. 704, pp. 1–14). American Chemical Society. https://doi.org/10.1021/bk-1998-0704.ch001

Tang, W.-W., Zeng, G.-M., Gong, J.-L., Liu, Y., Wang, X.-Y., Liu, Y.-Y., Liu, Z.-F., Chen, L., Zhang, X.-R., & Tu, D.-Z. (2012). Simultaneous adsorption of atrazine and Cu (II) from wastewater by magnetic multi-walled carbon nanotube. *Chemical Engineering Journal, 211–212*, 470–478. https://doi.org/10.1016/j.cej.2012.09.102

Thiruppathi, R., Mishra, S., Ganapathy, M., Padmanabhan, P., & Gulyás, B. (2017). Nanoparticle functionalization and its potentials for molecular imaging. *Advanced Science, 4*(3), 1600279. https://doi.org/10.1002/advs.201600279

Venkatesh, K., Arthanareeswaran, G., Chandra Bose, A., Suresh Kumar, P., & Kweon, J. (2021). Diethylenetriaminepentaacetic acid-functionalized multi-walled carbon nanotubes/titanium oxide-PVDF nanofiber membrane for effective separation of oil/water emulsion. *Separation and Purification Technology, 257*, 117926. https://doi.org/10.1016/j.seppur.2020.117926

Wang, K., Amin, K., An, Z., Cai, Z., Chen, H., Chen, H., Dong, Y., Feng, X., Fu, W., Gu, J., Han, Y., Hu, D., Hu, R., Huang, D., Huang, F., Huang, F., Huang, Y., Jin, J., Jin, X., . . . Zhong Tang, B. (2020). Advanced functional polymer materials. *Materials Chemistry Frontiers, 4*(7), 1803–1915. https://doi.org/10.1039/D0QM00025F

Wieszczycka, K., Staszak, K., Woźniak-Budych, M. J., Litowczenko, J., Maciejewska, B. M., & Jurga, S. (2021). Surface functionalization—the way for advanced applications of smart materials. *Coordination Chemistry Reviews, 436*, 213846. https://doi.org/10.1016/j.ccr.2021.213846

Wu, H., Sun, H., Hong, W., Mao, L., & Liu, Y. (2017). Improvement of polyamide thin film nanocomposite membrane assisted by tannic acid—FeIII functionalized multiwall carbon nanotubes. *ACS Applied Materials & Interfaces, 9*(37), 32255–32263. https://doi.org/10.1021/acsami.7b09680

Xin, Q., Wu, H., Jiang, Z., Li, Y., Wang, S., Li, Q., Li, X., Lu, X., Cao, X., & Yang, J. (2014). SPEEK/amine-functionalized TiO$_2$ submicrospheres mixed matrix membranes for CO$_2$ separation. *Journal of Membrane Science, 467*, 23–35. https://doi.org/10.1016/j.memsci.2014.04.048

Yassari, M., Shakeri, A., Salehi, H., & Razavi, S. R. (2022). Enhancement in forward osmosis performance of thin-film nanocomposite membrane using tannic acid-functionalized graphene oxide. *Journal of Polymer Research, 29*(2), 43. https://doi.org/10.1007/s10965-022-02894-x

# 3

## Preparation of Functional Nanomaterials and Polymers

### 3.1 Functionalization Methods

#### 3.1.1 Hydrothermal Method

The hydrothermal method uses water as a solvent in a closed system at a specific pressure and temperature to complete the reaction, simulating the formation of crystals during the natural mineralization process. Under hydrothermal circumstances, water characteristics such moisture content, density, surface tension, viscosity, and iconic product would be significantly changed. Crystals were created using the hydrothermal method in the early 1882.

| Homogeneous solution | → | Heat treatment process | → | Drying and washing | → | Nanoparticles and nanomaterials |

The system reaction temperatures can be significantly lowered, and highly crystalline products with restricted size distribution, high purity, and high aggregation can be produced. It is challenging to observe the material's growth because the reaction takes place in a closed system (Meng et al., 2016). The hydrothermal technique for crystal growth has the advantage of being able to produce crystalline phases that are unstable near the melting point. The hydrothermal technique can also be used to produce substances that are close to the melting point and have a high vapor pressure. The methodology is also ideal for growing huge, high-quality crystals while maintaining a tight control over their composition. The method's disadvantages include the requirement of expensive autoclaves and good quality seeds of a reasonable size, and the inability to observe the crystals as they grow. The hydrothermal method has another lot of drawbacks. The use of sealed pressure vessels makes it challenging to measure, understand, and control the reaction process. Additionally, only a small amount of the product is produced during each synthesis. A more approachable reaction would frequently result in a better comprehension and control of the reaction process. It is therefore not a preferred method for creating high-quality upconversion nanoparticles (UCNPs).

DOI: 10.12019781003391364-3

### 3.1.1.1 Synthesis of ZnS Nanoparticles by Hydrothermal Method

An autoclave, a thick-walled steel cylinder hermetically sealed to resist high temperatures and pressures for prolonged periods of time, is used for the hydrothermal process. A Teflon vessel is placed into the autoclave's interior cavity to ward off corrosion. The container has an inner diameter of 30 mm and a capacity of about 120 ml. Since every chemical utilized was of analytical grade, further purification was not necessary. We conducted tests with a range of $Zn^{2+}$ and $S^{2-}$ composition molar ratios at 220 °C. The details of the experiment were as follows: $ZnSO_4.5H_2O$ and $Na_2S.7H_2O$ powders were used as initial components. They were each dissolved separately in deionized water and agitated at room temperature for 30 minutes. Then, while stirring, droplets of the $Na_2S$ solution were added to the $ZnSO_4$ solution to mix them. The ratios of the sources of zinc and sulfur were 1:0.7, 1:1, and 1:1. Three molar ratios were employed to create samples 1, 2, and 3 in the synthesis process.

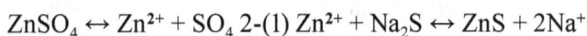

$$ZnSO_4 \leftrightarrow Zn^{2+} + SO_4 \, 2\text{-(1)} \, Zn^{2+} + Na_2S \leftrightarrow ZnS + 2Na^+$$

The Teflon-lined stainless steel autoclave was loaded with white gel before being filled to 50% capacity with the mixed solution. The sealed chamber maintained at 220°C for 12 hours in a box furnace (Mermert 500) before being allowed to cool naturally to ambient temperature. After centrifuging the product-containing solution, the supernatant layer was drained off and then washed with double-distilled water. The wash process was carried out ten times to completely eliminate contaminants from the sample. The final white product underwent a 12-hour, 60°C air-drying process. Using a Brucker D5005 diffractometer, X-ray diffraction was performed to investigate the structure of the ZnS samples. The microstructures and structures of the samples were characterized using a transmission electron microscope (JEOL JEM 1010) and a high-resolution transmission electron microscope (FEI Tecnai TF20 TEM/STEM). With the use of a FL3-22 Jobin Yvon Spex spectrofluorometer, the Photoluminescence (PL) and photoluminescence excitation (PLE) spectra were collected. The absorption spectra were obtained using a Shimadzu UV 2450 PC spectrometer. Diffuse reflectance was calculated using a UV-VIS-NIR spectroscope (Aneesh et al., 2007).

### 3.1.2 Solvothermal Method

Ferrite materials with enhanced physical (i.e., exact control over size distribution, shape, and crystalline phases) and chemical properties can be synthesized using the solvothermal process for use in industrial and biomedical applications. These physical features can be modified by adjusting certain experimental parameters, such as reaction temperature, reaction duration, solvent surfactant, and precursors. Solvothermal synthesis has been used to produce a number of ferrite materials and composites. Dissolution of chemical reagents and crystallization of products are aided by temperature and pressure conditions. For complex materials, this technology enables a one-step reaction strategy. High diffusivity is another benefit of using solvents, which increases the mobility of the dissolved ions and facilitates more effective reagent mixing. In the synthesis of inorganic compounds, aqueous solutions of simple salts, such as metal chlorides, nitrates, or acetates, are typically used as precursors (Nunes et al., 2019).

### 3.1.2.1 Solvothermal Synthesis of Hydrophobic and Hydrophilic Graphene Oxide Nanosheets

GO nanosheets were produced using natural graphite powder (45 m). Typically, powerful sulfuric acid and nitric acid solutions were stirred vigorously before adding natural graphite granules to the mixture. While the reaction vessel was immersed in an ice bath for 30 minutes, potassium chlorate was progressively added to the mixture. For 96 hours, the oxidation was let to continue. The oxidized suspension was centrifuged, washed with de-ionized water, and then washed with diluted (10 wt%) hydrochloric acid solution to remove the sulfate ions. The pH of the supernatant was increased numerous times until it reached 7. For exfoliating, an ultrasonic bath was employed. Graphene oxide particles were combined with distilled water. The fabrication of the GO nanosheets with allylamine successfully improved their dispersion in water. First, ethanol and allylamine were scattered onto GO nanosheets. Before being placed in an autoclave with a Teflon liner, the solution was stirred. The autoclave was heated to 90 °C and maintained for 0.5–2 hours to finish the reaction. The reaction product was rinsed once with 1:1 $H_2O$:EtOH and five times with acetone. At 60 °C, the final product was vacuum-dried.

N,N-dimethylformamide solution was used to dissolve the dried GO nanosheets. After that, a certain quantity of phenyl isocyanate was added. The mixture was then placed into a Teflon-lined autoclave and heated to 90 °C for 0.5–2 h to finish the reaction. The reaction result was then placed into a solvent—tetrahydrofuran (THF)—filtered, and repeatedly washed with THF to remove any residual phenyl isocyanate. The filtered product was then dried in a vacuum oven that was preheated to 60°C. The combination of phenyl isocyanate and functionalized GO nanosheets was then produced (G. Wang et al., 2009).

### 3.1.2.2 Synthesis of Reduced Graphene Oxide–$TiO_2$ Nanoparticles by Solvothermal Method

GO was made using the Hummers process. After being combined with 1.5 g of $NaNO_3$ in a cold bath, the graphite particles were added to concentrated $H_2SO_4$ (80 mL). Then, 10 g of $KMnO_4$ was added gradually, while being vigorously stirred, at a temperature below 20 °C. After that, the mixture was agitated for 3 hours at 35 °C in a water bath. After reaction, the mixture became pasty and brownish. The pasty mixture was placed in an ice bath to keep the temperature below 35 °C while 100 mL of water was gradually added. The solution's hue changed to a vivid yellow after 10 mL of 30% $H_2O_2$ solvent was added to the mixture. Then, 3 g of P25 powder was added after solid-state GO of various weights had been dissolved in

60 mL of glycol solution and the mixture had been sonicated for 2 hours to make it transparent. After vigorous stirring, the suspension was placed into a 100-mL Teflon-lined stainless-steel autoclave where it underwent a 6-hour reaction at 160 °C. The items were then vacuum-dried after being filtered with distilled water. The RGO-TiO₂ (P25) pastes are produced using the Grätzel technique. FTO conducting glass was submerged in a 40-mM aqueous $TiCl_4$ solution for 30 minutes at 70 °C in order to prepare the photo-anode films. Following this, the pastes were applied to the glass using a screen-printing technique (250 T mesh/inch, polyester). The films were then annealed in air for 30 minutes at 450 °C at a temperature rate of 5 °C/min. P25 films were prepared under the same conditions for comparison (Figure 3.1).

The spinning Pt counter electrode, N719 dye-sensitive RGO-TiO₂ composite sheets, and FTO conducting glass were assembled to form the solar cell, which was then sealed using Surlyn polymer frames. By dissolving 0.05 M I2, 0.5 M LiI, 0.3 M DMPII, and 0.5 M 4-TBP in acetonitrile, a typical redox electrolyte was formed (Shu et al., 2013).

### 3.1.3 Sol-Gel Method

A typical wet chemical technique to synthesize silica nanoparticles is the sol-gel approach. In the presence of a mineral acid or base as a catalyst, it involves the hydrolysis and condensation of inorganic salts like sodium silicate ($Na_2SiO_3$) or metal alkoxides like TEOS. There are a number of variables that affect the hydrolysis and condensation reactions (sol-gel process), including the activity of the metal alkoxide, the water/alkoxide ratio, solution pH, temperature, the kind of solvent used, and the additive. The sol-gel method can be applied in a variety of sectors, including the fabrication of ceramics and the transition between thin sheets of metal oxides (Bokov et al., 2021).

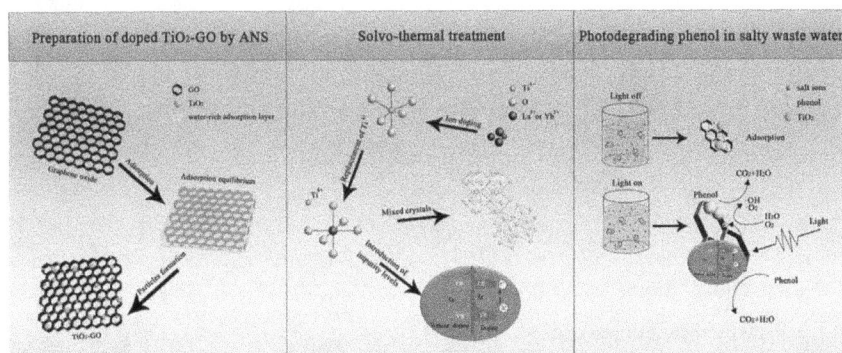

**FIGURE 3.1**

Functionalization of reduced GO with TiO₂ nanoparticles (T. Wang et al., 2019). *Reproduced with permission from Elsevier.*

## 3.2 Types of Polymerization

### 3.2.1 Anionic Polymerization

Anionic polymerization or Chain-growth polymerization is the process of polymerizing vinyl monomers and encompasses potent electronegative groups. It is a general method used to create synthetic polydiene rubber, solution styrene–butadiene rubber (SBR), and thermoplastic styrenic elastomers. Furthermore, some systems can only be polymerized by an anion mechanism in an aqueous solution, like cyclosiloxanes and cyclic ethers. Aqueous anionic polymerizations have been done in aqueous emulsions since the 1980s, despite their limitations, and their applications have grown significantly in significance. As a catalyst for the polymerization, a surfactant, the monomer, and a metal base were utilized in the earliest application of this technique. Alkaline Inisurfs, such asbenzyldimethyldodecylammonium hydroxide, were also used in these reactions. For the majority of monomers that can be polymerized using this process, the last set of conditions established constitutes a structure for anionic emulsion polymerization.

Broad molecular weight distributions (MWDs) and the formation of cyclic polysiloxanes are frequent as a result of the intricate reaction processes utilized in this procedure. Furthermore, numerous attempts have been made to carry out live anionic polymerization of siloxanes under various circumstances; nevertheless, these developments have been made in nonaqueous solvents. The derivatization and polymerization of anionic polymers (episulfides) are advantages of this process.

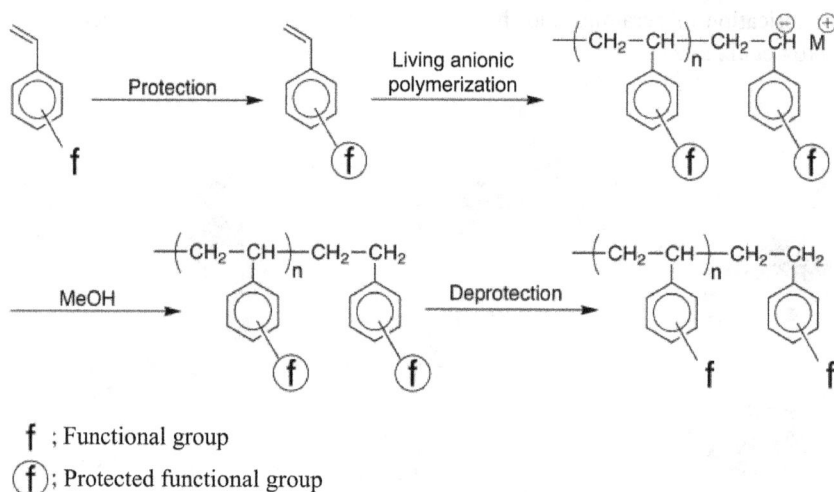

f ; Functional group

(f); Protected functional group

**SCHEME 3.1**
Living anionic polymerization of styrene (Hirao et al., 2002). *Reproduced with permission from Elsevier.*

Moreover, this is a sustainable green method for chemical and environmental engineering and science to produce NPs for drug delivery and other medicinal applications since poly(alkyl cyanoacrylates) are biodegradable.

### 3.2.1.1 Synthesis of Polymer/Silica Hybrid Nanoparticles Using Anionic Polymerization

Prior to usage, benzene and cyclohexane (Merck, g 99.7% and g 99.5%, respectively) were degassed and distilled over *n*-butyllithium. THF was degassed, pre-dried on CaH$_2$, and dried with Na/K and benzophenone before utilizing tetrahydrofuran (THF) (Merck, 99.9%). After being stirred over di-*n*-butylmagnesium for an entire night at 0 °C, degassed styrene (Fluka, g 99%) was distilled directly into the reactor.

Before use, butadiene (Aldrich, 99%) was directly condensed into the polymerization flask after being condensed on di-*n*-butylmagnesium and then on *n*-butyllithium and stirred at −20°C for 20 minutes. Tetrachlorosilane (SiCl$_4$), trimethylchlorosilane (Me$_3$SiCl), and dimethyldichlorosilane (Me$_2$SiCl$_2$) were distilled at the vacuum line (Aldrich, >99.5%, 99%, and 99%, respectively). Only the middle fraction, which included around one-third of the original substance, was used after the first fraction was removed. Toluene (Merck, g 99.9%), methanol (Merck, g 99.9%), *tert*-butyllithium, and *sec*-butyllithium were diluted in dry benzene or cyclohexane, and the precise concentration was determined using Gilman double titration. To be utilized as received, Nissan (NBAC-ST) provided silica NPs (R = 10 nm) in a 30% solution in *n*-butyl acetate (Hübner et al., 2010). The formation of surface-engineered zinc oxide NPs using a biocompatible PPEGMA polymer through free radical polymerization will be deliberately discussed.

### 3.2.2 Cationic Polymerization

Due to inter- or intramolecular transfer or rearrangement, cationic polymerization has long been considered an unreliable approach for making well-defined functional polymers. However, recently a number of monomers have been found to sufficiently control initiation, propagation, and termination to create functional polymers. For many years, cationic polymerization was infrequently used because THF was cationically polymerized using the difunctional initiator trifluoro anhydride. However, with the introduction of suitable catalysts and the right conditions, controlled living cationic polymerizations were achieved, resulting in better results even when using water as solvent. Using surfactants that might initiate the reaction (inisurfs), the earliest attempts to make polymers in an aqueous solution through cationic processes involved stabilizing monomer droplets in suspensions or emulsions.

Recently, well-controlled polymer designs have been synthesized from cationically polymerized vinyl ether monomers by methods resembling "living" polymerizations. These processes use relatively stable (i.e., less ionized) propagating species, effectively suppressing or completely preventing chain transfer. Consider the scenario when neither the solvent nor the strong nucleophiles are numerous. Assume that there is a sufficient amount of alkene. In this situation, the alkene may

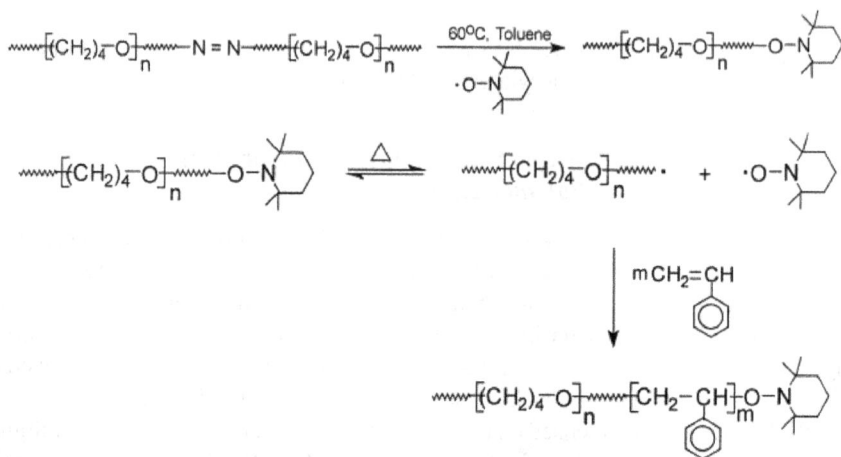

**SCHEME 3.2**
Cationic and stable mediated polymerization of block copolymers (Başkan Düz & Yağci, 1999). *Reproduced with permission from Elsevier.*

act as a nucleophile toward carbocation, developing a dimeric species with a new carbon–carbon bond and a new carbocation center.

Block copolymers with TEMPO-terminated polytetrahydrofuran and lead exhibit regulated sequence and narrow overall polydispersity during subsequent polymerization of styrene.

This type of polymerization is referred to as cationic polymerization because it involves a chain process that is propagated by carbocation intermediates. In order to minimize steric effects in the transition state (TS) for the addition process, cationic polymerization is especially effective for simple alkenes that generate moderately stable carbocations (notice the tertiary carbocation intermediate). As a result, isobutene is especially receptive to cationic polymerization, but other alkenes containing electron-donating groups (functionalities based on oxygen and nitrogen) are also effectively polymerized.

**Polymer Nomenclature:** Polymers can be named in a variety of ways. As shown above, for polyisobutylene, the simplest method is to put the name of the starting alkene (called the monomer) in parentheses and add the prefix poly.

## 3.2.3 Post Polymerization

Direct polymerization or copolymerization of monomers with chemoselective handles that are inert to the polymerization conditions but may be quantitatively changed into a broad spectrum of different functional groups in a subsequent phase is post-polymerization modification. The success of this method is based on the excellent conversions achieved under mild conditions, the excellent functional group tolerance, and the orthogonality of the post-polymerization modification reactions (Gauthier et al., 2009).

## (i) Plasma polymerization

Plasma polymerization integrates monomers with an energetic plasma species to form high-molecular-weight polymers from their constituent parts. Radicals on solid surfaces and/or gaseous monomers are produced by the plasma source, and these radicals are subsequently randomly recombined to create polymers. On a variety of surfaces, including metals and polymers, thin films can be made via plasma polymerization (Figure 3.2). Films that have been plasma-polymerized are smooth, uniform, and ultrathin (10 to a few hundred nanometers), with good adhesion and no pinholes. Plasma polymerization improves osseointegration by generating hydrophilic functional groups like amine ($-NH_2$) and carboxyl groups ($-COOH$), resulting in a beneficial surface modification approach (Gan & Berndt, 2015).

## (ii) Interfacial polymerization

Regulated fabrication of films, capsules, and fibers for use as separation membranes and electrode materials has offered substantial benefits for interfacial polymerization, which confines a chemical reaction at the liquid–liquid or liquid–air interface. Interfacial polymerization has been revitalized by recent developments in technology and polymer chemistry. The history of interfacial polymerization is examined in terms of polymerization processes such interfacial polyaddition, interfacial polycondensation, interfacial polycoordination, interfacial oxidative polymerization, and interfacial supramolecular polymerization. The challenges and potential offered by new methods for interfacial polymerization are highlighted, along with newly emerging functional materials. There is no doubt that interfacial polymerization will keep evolving and yield a variety of unique functional materials (Figure 3.3).

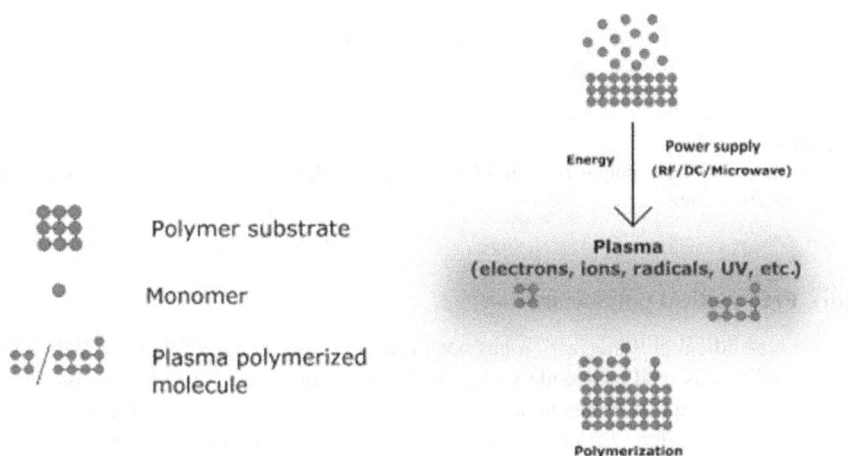

**FIGURE 3.2**
Mechanism of plasma polymerization (Nageswaran et al., 2019). *Reproduced with permission from Elsevier.*

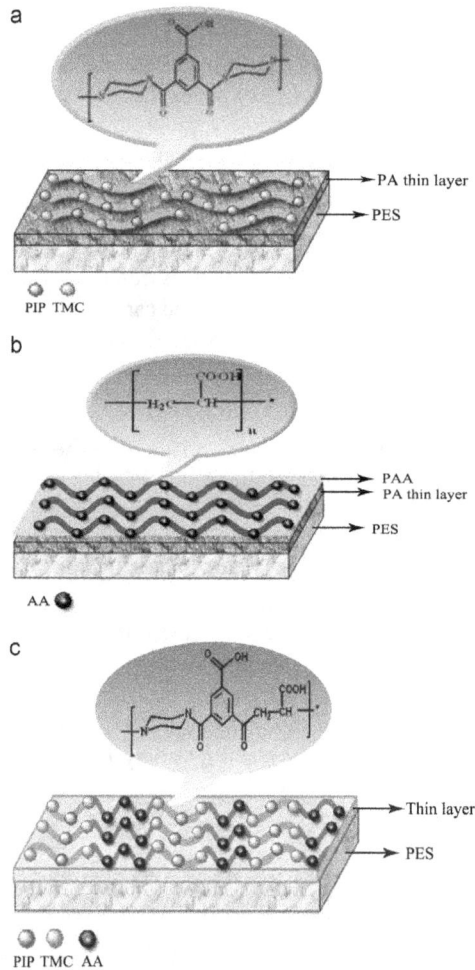

**FIGURE 3.3**
Mechanism of interfacial polymerization (Mansourpanah & Momeni Habili, 2013).*Reproduced with permission from Elsevier.*

(iii) **Free-radical polymerization**

Free-radical polymerization has been used to produce a notable number of commodities as well as a wide variety of lab-scale polymers. Free-radical polymerization, which tolerates polar functional groups, was the preferred technique for synthesizing many functional polymers. Free-radical polymerization has a number of drawbacks, including significant broad polydispersity and inadequate control over chain architecture and end groups. Telomerization is the procedure used when the chain transfer agent in free-radical polymerization is extremely reactive, and the resulting low molecular weights products are formed. The radical

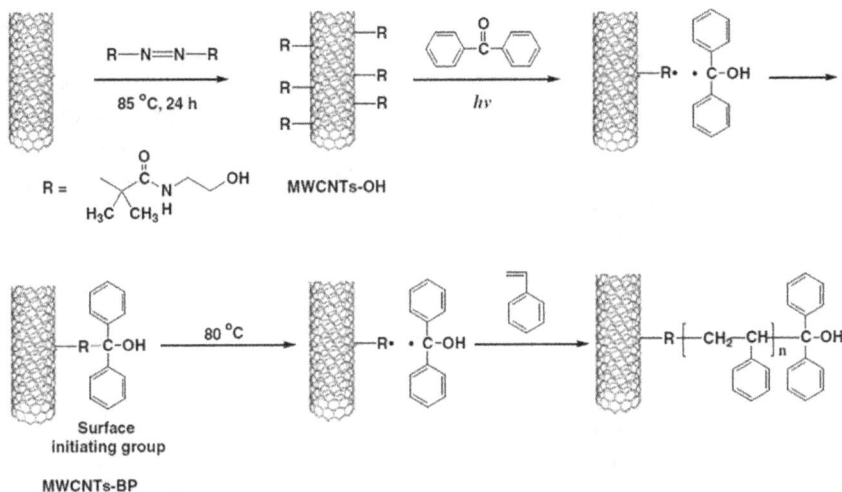

**FIGURE 3.4**
Mechanism of polymer-grafted MWCNTs (Park et al., 2010). *Reproduced with permission from Elsevier.*

polymerization mechanism for polymer-grafted multiwalled carbon nanotubes (MWCNTs) is illustrated in Figure 3.4.

### 3.2.3.1 Synthesis of PPEGMA-g-ZnO Nanocomposites Via in situ Free-radical Polymerization

The PPEGMA-g-ZnO nanocomposites were synthesized utilizing *in situ* free-radical polymerization in a single process. TMSPMA-f-ZnO NPs (1 g), PEGMA (5 ml), AIBN (0.05 g), and deionized water (10 ml) were added to a 100-mL round-bottom flask. After being subjected to ultrasonic vibrations for 30 minutes, the mixture was swirled in nitrogen for 12 hours at 65°C. After polymerization, the flask was cooled to room temperature, and the reaction mixture was precipitated in methanol. The product was filtered in a vacuum oven and dried for 24 hours (Islam et al., 2013).

Since peroxide O-O bonds are generally weak and rapidly break homolytically at mild thermal conditions, a peroxide is consistently employed as an initial catalyst when initiating a radical reaction.

### 3.2.4 Polycondensation

A polymer is created by linking one or more types of monomers to create long chains that release water or another simple component. This process is known as polycondensation. Aqueous polycondensation has developed significantly since its inception. Since ester, amide, and urethane linkages are all easily hydrolyzed, it may seem contradictory to polycondensation in aqueous circumstances, yet some polymerizations have been successfully completed utilizing both interfacial and emulsion polymerizations. This section offers a sample of typical reactions.

While efficient for producing these materials, this approach has drawbacks, such as the extra difficulty in sanitizing the final polymers and side effects brought on by high temperatures. A prominent alternative to this method is aqueous emulsion polycondensation, which produces polyesters. Aliphatic polyesters with nanometric particle sizes have been made using this approach.

As an illustration, consider the generation of peptides and proteins via the creation of amide bonds in solid–liquid interfacial polycondensation. By applying an amino acid to a solid substrate and leading it to react with another amino acid that has an activated carbonyl group and a protected amine group, these techniques can be used to produce proteins. The amine is then reacted and unprotected with an additional amino acid protected by an amine. For the creation of synthetic proteins, this process is essential. The formation of a lower molecular weight component during the polycondensation reaction typically affects the kinetics of polycondensation. The molecular component's mass and concentration will negatively impact the reaction process. The solution is to conduct the reaction at a higher temperature while keeping a strong vacuum in place, which will efficiently and effectively remove the by-products produced during the reaction and encourage the formation of larger molecular weight polymers. The most common polymers produced by the polycondensation reaction process include polyesters, nylons, and polyurethanes (Bhat & Kandagor, 2014).

### 3.2.5 *In Situ* Polymerization

*In situ* polymerization is a preparation method that occurs "in the polymerization mixture" and is used to create polymer nanocomposites from NPs. For utilization in various procedures, a large number of unstable oligomers (molecules) must be synthesized *in situ* (i.e., in the reaction mixture but cannot be isolated on their own). A hybrid formation of polymer molecules and NPs is created during the *in situ* polymerization process as a result of a series of polymerization phases that follow an initiation step.

Initially, liquid monomers or precursors with low molecular weight are used to disseminate NPs. The polymerization reaction is started by adding an appropriate initiator after forming a homogenous mixture, which is then exposed to heat, radiation, or another energy source. Following the completion of the polymerization mechanism, a nanocomposite made of polymer molecules bound to NPs is formed.

*In situ* polymerization of precursor polymer molecules is performed to form a polymer nanocomposite. The requirements for this process include the use of low viscosity pre-polymers (typically less than 1 pascal), a lesser time for polymerization, the use of polymers with advantageous mechanical properties, and additional product formation. Three different *in situ* polymerization procedures are distinguished by the functionality of the surface modifiers interacting with the monomers. The surface alteration simply produces a hydrophobic contact between the particles and the monomer, allowing for enhanced compatibility, if the amphiphilic

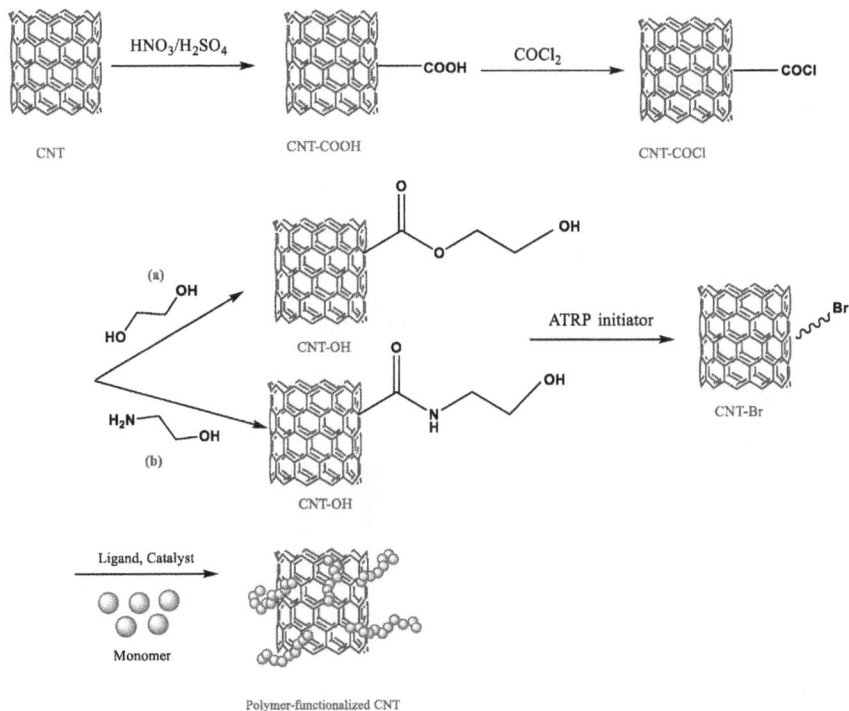

**FIGURE 3.5**
*In situ* polymerization mechanism of polymer functionalization of CNTs (Eskandari et al., 2021). *Reproduced with permission from Elsevier.*

ligand has simple alkyl chains. Depending on hydrophobicity of the monomer, correct chain lengths must be used to change the particle's hydrophobicity (Figure 3.5). For example, ligands with hydroxyl or carboxyl groups are preferable for dispersion in polar monomers because they enable higher conformance and hydrogen bonding to the monomers. Although the molecules of the surfactant are not directly involved in the polymerization process, monomers can dissolve into the hydrophobic surfactant layer and inorganic particles will become embedded in the developing polymer throughout the polymerization process (a process known as ad-polymerization) (K. Wang et al., 2020).

## 3.2.6 Synthesis of Functional Polymers by Post Modification

Polymerization of monomers with functional groups that are inert to the polymerization conditions but may be quantitatively transformed into a wide range of different functional groups in a subsequent reaction step is the basis for producing functional polymers by post-polymerization modification. In addition to providing access to functional polymers that cannot be made through direct polymerization of matching functional monomers, post-polymerization modification is a

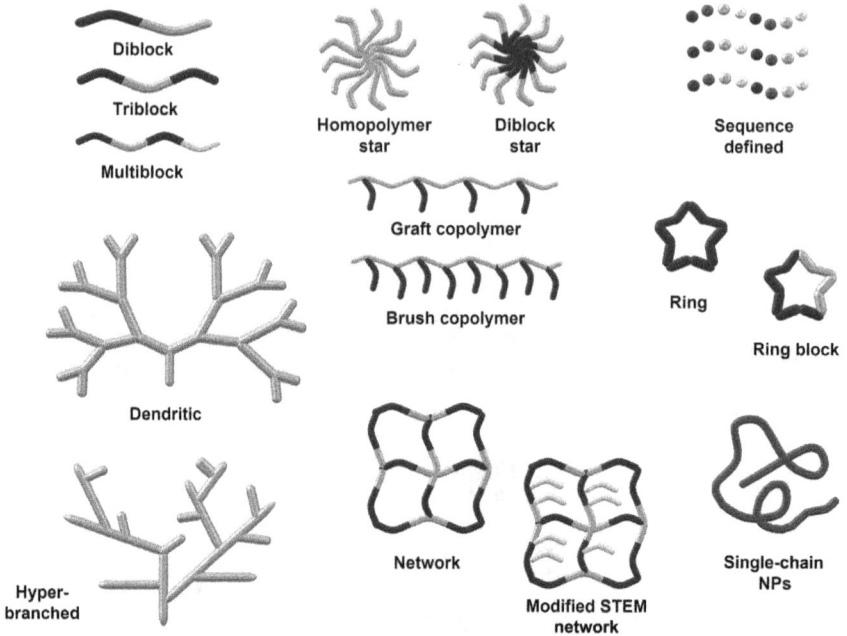

**FIGURE 3.6**
Functional polymer synthesis based on controlled polymerization methods (Corrigan et al., 2020). *Reproduced with permission from Elsevier.*

very attractive method for combinatorial material discovery. In order to generate a broad library of functional polymers with equal average chain length and chain-length distribution, a single reactive polymer precursor can be employed in the post-polymerization modification strategy. This substantially simplifies the process of establishing structure–property interactions (Gauthier et al., 2009).

### 3.2.6.1 Synthesis and Characterization of ZnO by TMSPMA

In a 500-mL three-neck round-bottom flask, 20 g of TMSPMA was added to a dispersion of 10 g of ZnO NPs in 200 mL of toluene. The resulting mixture was stirred under an argon gas atmosphere at room temperature for 20 hours. After reaction, TMSPMA-f-ZnO NPs were centrifuged and repeatedly washed with toluene. The powder was then vacuum-dried for 24 hours at 30°C.

## 3.3 Applications of Functional Polymers and Nanomaterials

Surface modification is necessary to improve adhesion, printing, and wetting by adding various types of polar and other functional groups on polymer and nanostructure surfaces. This is necessary because the inert nature of the majority of commercial polymers and NMs restricts their development for specific applications in various industries (Sharma et al., 2015). Many surface functionalization techniques have been developed over the past few decades, and they all generally follow a similar procedure: First, the primary reactive functional groups are bound to the polymer chain ends at the surface, and then the reactive surface is modified with active/bioactive agents, hydrophobic and hydrophilic monomers, oligomers, or polymers to achieve particular surface characteristics that correspond to the needs of the end use.

Immobilizing active/bioactive substances on polymeric surfaces frequently involves covalent bonding, electrostatic interactions, and ligand–receptor pairing. In some applications, including drug delivery systems, non-covalent physical adsorption is favorable. Covalent immobilizations also prevent bioactive substances from migrating from active food-packaging films to food, hence increasing the half-life of biomolecules and delaying their fast metabolism (Makvandi et al., 2021).

### 3.3.1 Wastewater Treatment

Industrial activities in recent decades have significantly increased the amount and variety of contaminant pollution in the aquatic environment, causing severe environmental damage.

A significant amount of contaminants in wastewater is harmful, and when they accumulate in living things, they threaten those things. These contaminants, including dyes, heavy metal ions, organic contaminants, and others, have been removed using various techniques and materials, including synthetic, natural, hybrid, waste, renewable, and engineered materials. Recent decades have seen a paradigm shift toward the use of NPs rather than bulk materials, leading to significant advancements in nanotechnology and the development of innovative NMs for various industrial and environmental applications. Due to the related physiochemical characteristics that the bulky phase could not carry, which made it possible for them to be utilized in many scientific domains, including water treatment, the focus on nanotechnology has changed. Researchers found that the inclusion of certain functional groups in functionalized polymers, as opposed to bare NPs, enabled efficient and selective adaptation to a particular pollutant. This section will begin by discussing how the physical condition of polymeric materials and the characteristics of functional moieties affect their use in a water treatment process. Therefore, the specific selectivity of polymer-functionalized materials containing different functional groups, such as carboxylic, hydroxyl, phosphonic, amine, and sulfonic, toward the targeted pollutant will be given special consideration.

### 3.3.2 Textile Applications

The textile business, which was first developed for human clothing but has subsequently been expanded with great success to provide several fascinating items in

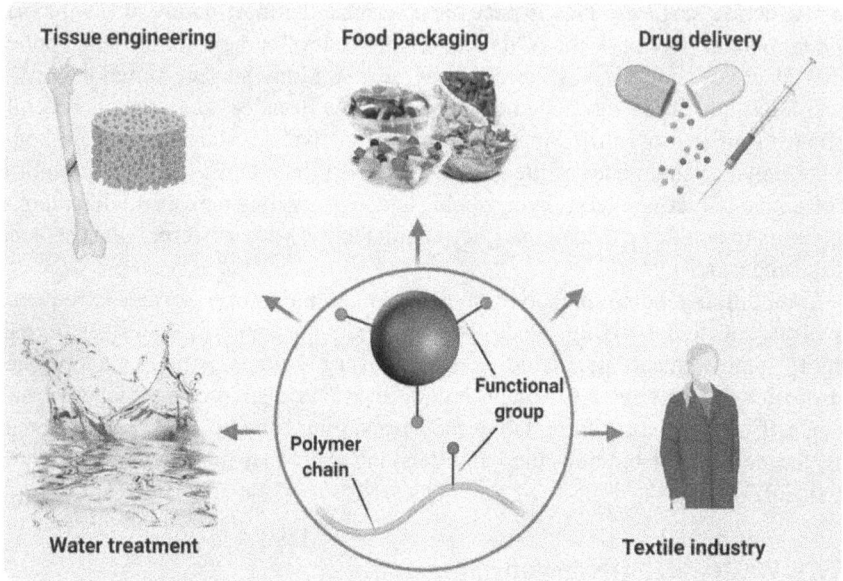

**FIGURE 3.7**
Various applications of functional polymers and NPs (Makvandi et al., 2021). *Reproduced with permission from Springer.*

recent years, is one of the most significant functional fields of polymer engineering. Years of research are needed to determine the optimum material for clothes for various environmental conditions, and a vast selection of products are now available that all aim to provide the most pleasant atmosphere for a wide range of consumers. Natural materials have been extensively used as solid-phase extraction absorbents for sample preparation due to their high adsorption capability. These eco-friendly materials possess desirable stability in nonaqueous or aqueous solutions, great mechanical strength, and acceptable biocompatibility. However, the monotonous functional groups of these materials limit their adaptability. Additionally, nanotechnology has made smart textiles possible thanks to the use of stimuli-responsive polymers that can detect and react to environmental changes. This development has the potential to provide a significant international market with an annual growth rate of roughly 26%.

Based on the induced properties, the surface-functionalized texture via different agents leads to the introduction of various types of smart textile that can be classified into various categories, some of which are as follows:

1. Textiles with antimicrobial properties

2. Self-cleaning textiles

3. Moisture-wicking textiles and cooling management

4. Textiles that are flame retardant

5. Textiles that heal themselves

### 3.3.3 Food Packaging

Polymer nanoparticles (PNP) use in food packaging has increased significantly due to their advantages over traditional materials. Polymer-functionalized materials tend to be developed to provide strength, heat resistance, stability, and stiffness as well as to enhance UV ray and gas barrier attributes—one explanation why the biggest food corporations in the world are investing money on research to develop packaging materials with better barrier, mechanical, and antibacterial characteristics. PNP applications will be divided into three subcategories in this context: enhanced, active, and intelligent PNP packaging materials.

### 3.3.4 Other Industrial Applications

Polymer-functionalized NMs are part of the polymer nanotechnology spectrum, which is widely used in an interdisciplinary field. This section provides an overview of polymer-functionalized materials' application domains in sensors, energy storage, and catalysis.

### 3.3.5 Sensors

The chemical and structural changes brought about by the interaction of the polymer and NPs with the environment control the specific focusing of PNPs. This results in a subsequent output, which is typically in the form of an optical or electrical signal in the presence of vapors of toluene and benzene at room temperature with polyaniline metal oxide composites ($TiO_2/SnO_2$). Surface adsorption happens, and the sensor material interacts with analyte vapors.

It has recently been discovered that the combination of CNTs and polymer can interact with specific molecules. This interaction has aided in increasing the strength of electrochemical signals, resulting in the creation of sensitive and selective sensors. Pyrrole is a typical polymer for surface modification of CNTs as a monomer and is suitable for biological and pharmacological analyses ranging from acidic to neutral because of its conductive characteristics.

### 3.3.6 Energy Storage

The development of renewable and sustainable storage systems, such as wind and solar energy, is urgently required in light of the growing energy crisis brought on by the depletion of conventional fossil fuels. Effective energy storage systems are necessary to produce and store renewable energy rapidly and steadily since renewable energy resources are intermittent. Cells, batteries, and capacitors are a few of the different storage methods that have been crucial, and this has led to a rise in research interest in electrochemical energy storage systems in general.

Lithium-ion batteries, which are built on four basic components consisting of an anode, a cathode, a separator, and an electrolyte, are a suitable application for PNPs. Presently, porous polyolefin-based polymers, such as polyethylene and polypropylene, are used in separators that limit electrolyte diffusion due to their hydrophobic surfaces.

### 3.3.7 Catalysis

Catalytic activity is present in a wide range of PNP-type nanoreactors, for instance, metal NPs that are chemically drawn to polymer chains, or the uniform distribution of functionalized NPs in microgel assemblies, on polymeric membrane surfaces, or in metallic thin films that have been functionalized with polymers. For hydrogen evolution, functionalized polymer brushes are developed. The cationic polyelectrolyte brushes, which catalyze the formation of hydrogen from water by binding molybdenum sulfide of cationic polyelectrolytes, were produced using highly oriented planar pyrolytic graphite.

### References

Aneesh, P. M., M, A., & Jayaraj, M. (2007). Synthesis of ZnO nanoparticles by hydrothermal method. *Nanophotonic Mater IV*. https://doi.org/10.1117/12.730364

Başkan Düz, A., & Yağci, Y. (1999). Synthesis of block copolymers by combination of atom transfer radical and promoted cationic polymerization mechanisms. *European Polymer Journal, 35*(11), 2031–2038. https://doi.org/10.1016/S0014-3057(99)00006-3

Bhat, G., & Kandagor, V. (2014). 1—Synthetic polymer fibers and their processing requirements. In D. Zhang (Ed.), *Advances in filament yarn spinning of textiles and polymers* (pp. 3–30). Woodhead Publishing. https://doi.org/10.1533/9780857099174.1.3

Bokov, D., Turki Jalil, A., Chupradit, S., Suksatan, W., Javed Ansari, M., Shewael, I. H., Valiev, G. H., & Kianfar, E. (2021). Nanomaterial by sol-gel method: Synthesis and application. *Advances in Materials Science and Engineering, 2021*, e5102014. https://doi.org/10.1155/2021/5102014

Corrigan, N., Jung, K., Moad, G., Hawker, C. J., Matyjaszewski, K., & Boyer, C. (2020). Reversible-deactivation radical polymerization (controlled/living radical polymerization): From discovery to materials design and applications. *Progress in Polymer Science, 111*, 101311. https://doi.org/10.1016/j.progpolymsci.2020.101311

Eskandari, P., Abousalman-Rezvani, Z., Roghani-Mamaqani, H., & Salami-Kalajahi, M. (2021). Polymer-functionalization of carbon nanotube by in situ conventional and controlled radical polymerizations. *Advances in Colloid and Interface Science, 294*, 102471. https://doi.org/10.1016/j.cis.2021.102471

Gan, J. A., & Berndt, C. C. (2015). 4—Plasma surface modification of metallic biomaterials. In C. Wen (Ed.), *Surface coating and modification of metallic biomaterials* (pp. 103–157). Woodhead Publishing. https://doi.org/10.1016/B978-1-78242-303-4.00004-1

Gauthier, M. A., Gibson, M. I., & Klok, H.-A. (2009). Synthesis of functional polymers by post-polymerization modification. *Angewandte Chemie International Edition, 48*(1), 48–58. https://doi.org/10.1002/anie.200801951

Hirao, A., Loykulnant, S., & Ishizone, T. (2002). Recent advance in living anionic polymerization of functionalized styrene derivatives. *Progress in Polymer Science, 27*(8), 1399–1471. https://doi.org/10.1016/S0079-6700(02)00016-3

Hübner, E., Allgaier, J., Meyer, M., Stellbrink, J., Pyckhout-Hintzen, W., & Richter, D. (2010). Synthesis of polymer/silica hybrid nanoparticles using anionic polymerization techniques. *Macromolecules, 43*(2), 856–867. https://doi.org/10.1021/ma902213p

Islam, Md. R., Bach, L. G., Jung, S.-J., Gal, Y.-S., & Lim, K. T. (2013). Surface engineering of zinc oxide nanoparticles by biocompatible PPEGMA polymer: Synthesis, characterization, and optical property studies. *Molecular Crystals and Liquid Crystals, 580*(1), 39–46. https://doi.org/10.1080/15421406.2013.803912

Makvandi, P., Iftekhar, S., Pizzetti, F., Zarepour, A., Zare, E. N., Ashrafizadeh, M., Agarwal, T., Padil, V. V. T., Mohammadinejad, R., Sillanpaa, M., Maiti, T. K., Perale, G., Zarrabi, A., & Rossi, F. (2021). Functionalization of polymers and nanomaterials for water treatment, food packaging, textile and biomedical applications: A review. *Environmental Chemistry Letters, 19*(1), 583–611. https://doi.org/10.1007/s10311-020-01089-4

Mansourpanah, Y., & Momeni Habili, E. (2013). Preparation and modification of thin film PA membranes with improved antifouling property using acrylic acid and UV irradiation. *Journal of Membrane Science, 430*, 158–166. https://doi.org/10.1016/j.memsci.2012.11.065

Meng, L.-Y., Wang, B., Ma, M.-G., & Lin, K.-L. (2016). The progress of microwave-assisted hydrothermal method in the synthesis of functional nanomaterials. *Materials Today Chemistry, 1–2*, 63–83. https://doi.org/10.1016/j.mtchem.2016.11.003

Nageswaran, G., Jothi, L., & Jagannathan, S. (2019). Chapter 4—Plasma assisted polymer modifications. In S. Thomas, M. Mozetič, U. Cvelbar, P. Špatenka, & K. M. Praveen (Eds.), *Non-thermal plasma technology for polymeric materials* (pp. 95–127). Elsevier. https://doi.org/10.1016/B978-0-12-813152-7.00004-4

Nunes, D., Pimentel, A., Santos, L., Barquinha, P., Pereira, L., Fortunato, E., & Martins, R. (2019). 2—Synthesis, design, and morphology of metal oxide nanostructures. In D. Nunes, A. Pimentel, L. Santos, P. Barquinha, L. Pereira, E. Fortunato, & R. Martins (Eds.), *Metal oxide nanostructures* (pp. 21–57). Elsevier. https://doi.org/10.1016/B978-0-12-811512-1.00002-3

Park, J. J., Park, D. M., Youk, J. H., Yu, W.-R., & Lee, J. (2010). Functionalization of multiwalled carbon nanotubes by free radical graft polymerization initiated from photoinduced surface groups. *Carbon, 48*(10), 2899–2905. https://doi.org/10.1016/j.carbon.2010.04.024

Sharma, R. K., Sharma, S., Dutta, S., Zboril, R., & Gawande, M. B. (2015). Silica-nanosphere-based organic—inorganic hybrid nanomaterials: Synthesis, functionalization and applications in catalysis. *Green Chemistry, 17*(6), 3207–3230. https://doi.org/10.1039/C5GC00381D

Shu, W., Liu, Y., Peng, Z., Chen, K., Zhang, C., & Chen, W. (2013). Synthesis and photovoltaic performance of reduced graphene oxide—TiO2 nanoparticles composites by solvothermal method. *Journal of Alloys and Compounds, 563*, 229–233. https://doi.org/10.1016/j.jallcom.2013.02.086

Wang, G., Wang, B., Park, J., Yang, J., Shen, X., & Yao, J. (2009). Synthesis of enhanced hydrophilic and hydrophobic graphene oxide nanosheets by a solvothermal method. *Carbon, 47*(1), 68–72. https://doi.org/10.1016/j.carbon.2008.09.002

Wang, K., Amin, K., An, Z., Cai, Z., Chen, H., Chen, H., Dong, Y., Feng, X., Fu, W., Gu, J., Han, Y., Hu, D., Hu, R., Huang, D., Huang, F., Huang, F., Huang, Y., Jin, J., Jin, X., . . . Zhong Tang, B. (2020). Advanced functional polymer materials. *Materials Chemistry Frontiers, 4*(7), 1803–1915. https://doi.org/10.1039/D0QM00025F

Wang, T., Li, B., Wu, L., Yin, Y., Jiang, B., & Lou, J. (2019). Enhanced performance of TiO2/ reduced graphene oxide doped by rare-earth ions for degrading phenol in seawater excited by weak visible light. *Advanced Powder Technology, 30*(9), 1920–1931. https:// doi.org/10.1016/j.apt.2019.06.011

# 4

## Structural Properties of Functional Nanomaterials and Polymers

### 4.1 Overview of the Structure and Properties of Functional Nanomaterials and Polymers

Functional groups are collections of one or more atoms that exhibit recognizable chemical properties regardless of what is connected to them. Functional group atoms are bonded to one another and to other molecules by covalent bonds. The functional groups known as ligands interact with and bind to the central atom of a coordination complex.

A functional group is defined as a "specific grouping of elements for which the peculiar chemical reactions of these molecules are responsible." Similar or identical chemical reactions will occur when two molecules of differing sizes but with the same functional groups are involved. When a molecule has a functional group, this indicates that one may predictably and systemically anticipate its behaviors and chemical interactions. Chemical synthesis—the deliberate carrying out of chemical reactions to produce a specific compound—can be produced by understanding the features of various functional groups.

Functional groups are atomic assemblages that are connected to the carbon-based core of organic components. Functional groupings are the end result of unique chemical processes involving organic compounds. They are less stable and more prone to participate in chemical reactions than the carbon backbone.

Some of the crucial functional groups in biological compounds are the hydroxyl, carbonyl, methyl, amino, carboxyl, phosphate, and sulfonic groups (Figure 4.1).

### 4.2 Structures and Properties

#### 4.2.1 Amine Group-based Functionalization of Nanomaterials and Polymers

Amines are organic substances with the functional group $-NH_2$, which contains a single electron and a nitrogen atom. These are ammonia derivatives in which aryl or

DOI: 10.1201/9781003391364-4

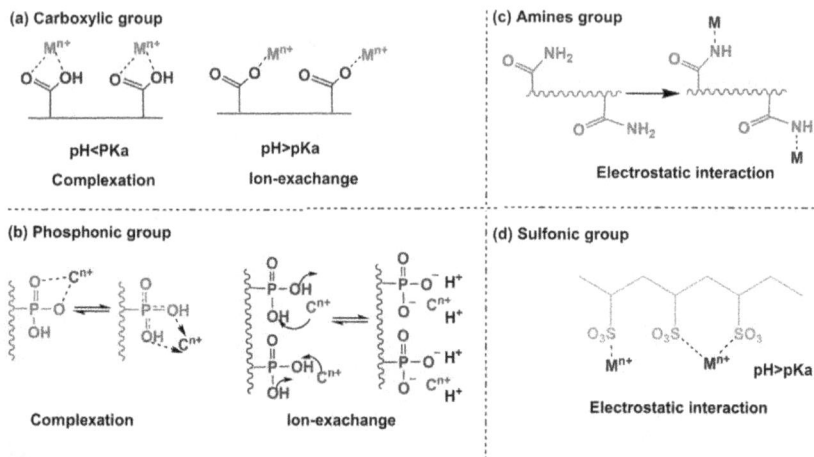

**FIGURE 4.1**

Functionalization mechanism of various functional groups (Makvandi et al., 2021). *Reproduced with permission from Springer.*

alkyl groups can take the place of one or more hydrogen atoms. Significant amines include amino acids, biogenic amines, trimethylamine, and aniline. Various amine-based chemicals are widely employed in the post-combustion $CO_2$ capture process (CCP) at natural gas conditioning plants to remove acid gas pollutants, and in energy storage and water treatment applications such as ethanolamine (EA), monoethanolamine (MEA), N-methyldiethanolamine (MDEA), and diethanolamine (DEA). The environment and human health are significantly impacted by these amines.

Lower aliphatic amines are gaseous in nature and have a fishy odor. Primary amines with three or four carbon atoms are liquid at room temperature, but those with more atoms are solid. Normally colorless, aniline and other arylamines oxidize in the air when stored outside, resulting in their color. Some examples of organic solvents that quickly dissolve amines are alcohol, benzene, and ether. Alcohol is more polar than amines, which results in stronger intermolecular hydrogen bonding.

Because of hydrogen bonds formed by the nitrogen of one molecule and the hydrogen of the other, primary and secondary amines frequently interact with one another. Organic substances known as amines frequently rely on nitrogen atoms and contain one or more of them. Although amines are structurally like ammonia, in that up to three hydrogen atoms may be bonded by nitrogen, they also possess additional qualities that are reliant on their connectedness to carbon (Mukherjee et al., 2019; Salim, 2021).

Amines are substances that have at least one amino group as their functional group. Amines are weak bases, which create when a nitrogen atom that undergoes sp3 hybridization and bonds to one or more carbon atoms. At room temperature, nitrogen hybridization shifts from sp3 to sp2 in a transition state, causing amine inversion to occur quickly. Amines are divided into three categories: primary ($RNH_2$), secondary (RRNH), and tertiary (RRRN) amines, depending on how many carbon atoms are linked to the hybrid nitrogen atom. On the other hand, a polymer is a lengthy chain of monomers, which are chemical units that can repeat. The longest chain of atoms in a continuous molecule that are covalently bonded is the polymer backbone. Amines are a component of some polymers, which are referred to as "amine-rich polymers" (ARPs).

The source components of the ARPs were categorized as either natural or synthetic. The chemical structure and key synthesis pathways have been demonstrated weakly. Next, the significance of amine groups for pollutant adsorption, flocculation/coagulation, photocatalysis, and membrane purification has been investigated. The transformation of polymer amine groups into quaternary groups is addressed in Table 4.1.

In heterogenized amines, nitrogen is linked to two hydrogen atoms, alkyl groups, or a combination of these groups. The nitrogen atom is surrounded by an electron density region and one unpaired electron. There is a partial negative charge because of the nitrogen's electronegativity and electron density. As a result, the acid medium's hydrogen ion can associate with the N, resulting in a positive charge. Due to its cationic composition, it can electrostatically absorb negatively charged organic and inorganic pollutants. The pH of the fluid and the surface zeta-potential both have an impact on cationic behavior. According to Scheme 4.1, the surface becomes more charged as the pH decreases toward the isoelectric point and vice versa. The phenomenon cannot be explained by electrostatic attraction (Dong et al., 2022; Elhalwagy et al., 2023).

### 4.2.2 Imine Group-based Functionalization of Nanomaterials and Polymers

A functional group or chemical molecule with a carbon–nitrogen double bond is known as an imine. Imines are chemical substances with a double bond between carbon and nitrogen (C=N). The oxygen atoms of aldehydes and ketones are replaced with the (N-R) group to create imines Their general formula is R2C=NR. Chemical molecules with nitrogen include imines and enamines. Imines are organic chemicals with a C=N functional group, whereas enamines are organic substances with an amine group near to a C=C double bond. The primary difference between imines and enamines is that imines have a C=N bond while enamines have a C-N bond.

Imines are created when a primary amine (or ammonia) combines with a carbonyl-containing acid ($H^+$, $H_3O^+$), and they resemble carbonyls. They resemble carbonyls in their purest form, but with a double bond to nitrogen instead of oxygen. When the reaction is complete, we can see a structural resemblance. In the infrared spectrum, oximes contain roughly three characteristic bands with wave numbers measuring 3600 (O-H), 945 (N-O), and 1665 (C=N) cm$^{-1}$. Compared to similar hydrazones, the aliphatic group of oximes is more resistant to the hydrolysis process. These substances are reported to be less soluble in water and are found to be colorless crystals. Oximes are thought to be poisonous by nature and to have weak acidic and basic characteristics. The temperature of these compounds is

**TABLE 4.1**

Illustrative examples of dye adsorbent nanomaterials based on magnetite and ARPs (Elhalwagy et al., 2023). *Reproduced with permission from Elsevier.*

| Adsorbent | Pollutant | Type | $q_e$, max (mg/g) | Time (min) | Optimum condition | | | | Ref. |
|---|---|---|---|---|---|---|---|---|---|
| | | | | | Adsorbent dose (g/l) | Pollutant dose (mg/l) | Temp. (°C) | pH | |
| Fe$_3$O$_4$/CS | Methyl orange | Anionic | 638.6 | 360 | 0.1 | 200 | 29.85 | 4 | (B. Chen et al., 2020) |
| Fe$_3$O$_4$/CS | Methylene blue | Cationic | 0 | — | — | — | — | — | (B. Chen et al., 2020) |
| Fe$_3$O$_4$/poly-lysine | Methyl blue | Anionic | 318.47 | 30 | 1 | 200 | — | 2 | (Y.-R. Zhang et al., 2015) |
| Fe$_3$O$_4$/PDA | Methylene blue | Cationic | 10 | 360 | 0.5 | 6 | — | 7 | (Zhou & Liu, 2017) |
| Fe$_3$O$_4$/PDA | Methyl orange | Anionic | 2.97 | 360 | 0.5 | 6 | — | 2 | (Zhou & Liu, 2017) |
| Fe$_3$O$_4$/PANi | Basic blue 3 | Cationic | 78.13 | 60 | 0.1 | 50 | 30 | 10 | (Muhammad, Shah, Bilal et al., 2019) |
| Fe$_3$O$_4$/PANi | Acid blue 40 | Anionic | 216.9 | 50 | 0.1 | 100 | 30 | 5.5 | (Muhammad, Shah, & Bilal, 2019) |
| Fe$_3$O$_4$/PPy | Brilliant green | Cationic | 263.85 | 20 | 1 | 320 | 30 | — | (M. Zhang et al., 2020) |
| Fe$_3$O$_4$/PPy | Eosin Y | Anionic | 226.76 | 20 | 1 | 320 | 35 | — | (M. Zhang et al., 2020) |
| Fe$_3$O$_4$/PPy | Methyl orange | Anionic | 152.91 | 45 | 1 | 320 | 35 | — | (M. Zhang et al., 2020) |
| Fe$_3$O$_4$/PEI | Crystal violet | Cationic | 372.6 | 90 | 1 | 700 | 24.85 | 8 | (Wan et al., 2022) |
| Fe$_3$O$_4$/PEI | Congo red | Anionic | 325 | 10 | 1 | 350 | 25 | 3 | (Liu et al., 2022) |

impacted by acid salts. Oximes prefer to disintegrate at higher temperatures, which might cause a huge explosion.

Imine compounds are a class of molecules with the imine group (-HC=N) and are often produced by combining an active carbonyl group from aldehydes or ketones with the amino group in primary amines. These adaptable precursors, which have a variety of pharmacological applications and biological functions, are used in the ring-closure synthesis of commercial chemicals. When perchloric acid is present, 4-fluorobenzaldehyde and 1-benzylpiperidin-4-amine successfully react to produce the imine product.

In multicomponent reactions, imines display an intriguing chemical behavior. Figure 4.2 displays the many imine reactivities that might be feasible. An imine's nitrogen atom, which is rich in electrons, might function as a nucleophile. Imines, on the other hand, behave as masked carbonyls/electrophiles and produce Mannich-type products when there are good C-nucleophiles present. Imines can become more electrophilic by protonating the nitrogen atom with a Bronsted acid. Imines can also be employed in cycloaddition reactions as dienophiles or even azadiene. Imines have a wide range of potential reactions, yet they frequently respond in

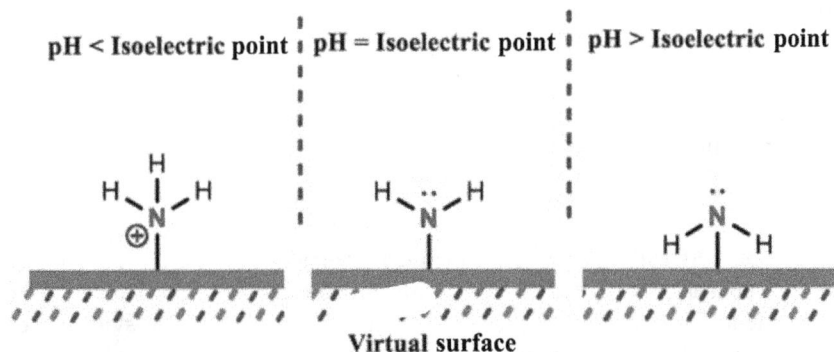

**SCHEME 4.1**
Response of surface amine groups at various pH levels based on the isoelectric point (Elhalwagy et al., 2023). *Reproduced with permission from Elsevier.*

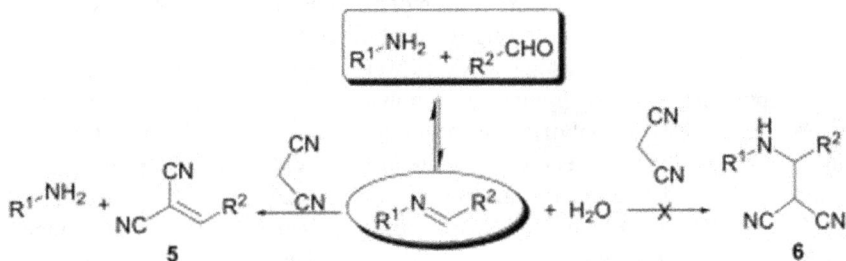

**FIGURE 4.2**
Mannich imine reaction (Choudhury & Parvin, 2011). *Reproduced with permission from Elsevier.*

certain situations with amazing selectivity. An imine's reactivity in a certain multicomponent reaction depends on both the nature of the reaction partner and the substituents of the imine.

The number of oxime groups in the reactivator's molecule is the sixth structural prerequisite that will be discussed in this section. One or more oxime groups with one, two, or three quaternary nitrogens can be present in an AChE reactivator molecule. Several structures with one, two, three, or four oxime groups are examples.

As can be observed, the prospective AChE reactivator potency is not increased by the presence of additional oxime groups. This finding might be related to the AChE reactivator's larger molecule size. The second oxime group's existence is not strictly required because the first oxime group (which has the lowest pKa) plays the primary role in the reactivation process.

The versatile molecule known as hydrazone belongs to the organic class and has the basic formula R1R2C=NNR3R4. Although hydrazone contains two nucleophilic nitrogen atoms, the reactiveness of nitrogen of the amino type is higher. The carbon atom, in contrast, possesses both nucleophilic and electrophilic characteristics. A class of chemicals having the formula R1R2C=NNH2 is known as hydrazones. Due to the substitution of the functional group $=NNH_2$ for the oxygen atom $=O$, they are connected to ketones and aldehydes. They typically result from the reaction of hydrazines with ketones or aldehydes (Abid et al., 2017; Choudhury & Parvin, 2011).

### 4.2.3 Phosphonic Group-based Functionalization of Nanomaterials and Polymers

Phosphorus-containing polymers have become increasingly important in recent years due to their intriguing properties that make them appropriate for a wide range of applications. As carbon-based polymers are stretched to their limits, searching for novel properties that are not possible with carbon can lead to introducing main group elements into polymer backbones (Vidal & Jäkle, 2019). The synthesis of a large number of stable monomers and polymers with distinctive and interesting properties, such as improved organo-solubility, good thermal and mechanical stabilities, modified flame retardancy, refractive index, gas separation ability, and improved proton transportation in hydrated conditions, has been made possible by the various chemical environments in which phosphorus atoms can exist (Fu et al., 2010; Kim et al., 2003). The incorporation of phosphorus into polymeric chains has been conveyed over time via various covalent linkages.

These phosphorus-containing linkages were integrated in the polymer chains of homopolymers and copolymers with polyamide, polyimide, polyphosphonate, poly(arylene ether), and polytriazole backbones using various polymerization techniques. The use of polymers as flame retardants, proton exchange membranes, adhesives, medical compounds, and optical materials has changed as a result of the presence of phosphorus in polymers. According to Bian et al. (2010) and Salmeia & Gaan (2015), a significant amount of phosphorus-based polymers are employed as flame-retardant materials. Proton exchange membranes for fuel cell applications have shown promise in a number of phosphine oxide groups incorporating sulfonated aromatic polymers (Mandal et al., 2019). Gas separation applications have also made use of phosphorus-containing polyamide and polyimide membranes (Bisoi et al., 2017). Additionally, the phosphorus-containing polymers' tunable optoelectronic properties led to their development of a high refractive index and a very effective blue electrophosphorescence (Shao et al., 2012a).

This comprehensive discussion aims to offer some general overview on the following subjects: (i) the synthesis of different phosphorus-based monomers via various chemical routes, (ii) polymerization of the synthesized monomers along with a few commercially available monomers, (iii) structural modifications of the polymer chains to achieve the desired properties, and (iv) membrane-based applications of phosphorus-containing aromatic polymers. On the other hand, Wehbi et al. (2020) and Monge et al. (2011) have previously provided an extensive overview of aliphatic-based phosphorus-containing polymers and their applications in the biomedical field.

Later, many phosphorus-containing polymers with varied polymeric backbones were synthesized and used in various applications. Polyphosphazines (Allcock & Kugel, 2002), polyesters (Stackman, 2002), polyamides (Bisoi et al., 2017), polyphosphonites (Petrov et al., 1963), polyurethanes (Chang et al., 1995), polyimides (Sato et al., 1980), poly(arylene ether)s (Smith et al., 1991), polycarbonates (Shan Wang & Yueh Shieh, 1999), polytriazoles (Ghorai et al., 2020), and others are among these polymers. High-performance phosphorus-based polymers, such as poly(arylene ether)s, polyimides, polyamides, polytriazoles, and others have been recognized as prospects in a variety of specialized applications due to their excellent thermal, mechanical, chemical, and flame-retardant properties.

Ghorai et al. (2020) examined the proton exchange membrane (PEM) properties of different phosphorus-containing sulfonated polytriazoles with high proton conductivity synthesized recently via click polymerization. Because of their simple synthesis routes and adaptability to various chemical designs, aromatic polymers with acidic functions are regarded as one of the promising candidates for proton exchange membranes (Hickner et al., 2004). Poly(arylene ether ketone)s (Miyake et al., 2013), polyphenylenes (Umezawa et al., 2012), polytriazoles (Ghorai et al., 2020), polyimides (Mandal et al., 2019), poly(arylene ether sulfone)s (Fu et al., 2010), and polybenzimidazoles (Jespersen et al., 2009) are examples of common aromatic polymers (Table 4.2). According to a recent study, multiblock copolymers with sulfonated hydrophilic and non-sulfonated hydrophobic components exhibit noticeably superior properties (Elabd & Hickner, 2011). Bae et al. (2010) reported on a number of very dense sulfonated aromatic polymers for PEM applications. The remarkable hydrophilic-hydrophobic phase-separated morphology with interconnecting ionic channels is the cause of the high proton conductivity. Oxidative stability is a further essential element for the operation and effectiveness of any polymer electrolyte membrane fuel cell (PEMFC). The inclusion of any electron-withdrawing chemical linkage (sulfone, ketone, and phosphine oxide moiety) in the polymeric structures, in addition to the hydrophilic part, has improved the oxidative stability of the membranes. The hydrophilic segments in the polymer chain significantly impact peroxide stability (Miyake et al., 2013).

**TABLE 4.2**

Phosphorus-containing polymers for optoelectronic applications (Ghorai & Banerjee, 2023). *Reproduced with permission from Elsevier.*

| Polymeric backbone | Chemical structure | Features | Refs. |
|---|---|---|---|
| Polyphosphonates | | High refractive index values (1.58–1.60) | Shobha et al. (2001) |
| Polyimides | | Good adhesive property (104.7–126.3 g/mm), and very low dielectric constant (2.34–2.89) | Choi et al. (2009) |
| Poly(arylene ether phosphine oxide) | | Bipolar polymeric host for highly efficient blue electro-phosphorescence | Shao et al. (2011) |

| | | | |
|---|---|---|---|
| Fluorinated poly(arylene ether phosphine oxide) | | Highly efficient blue electro-phosphorescent polymers | Shao et al. (2012a) |
| Poly(arylene ether phosphine oxide) | | Highly efficient all-phosphorescent single white-emitting polymers | Shao et al. (2012b) |
| Poly(phenylcarbazole-alt-triphenylphosphine oxide) siloxane | | Bipolar host material for solution-processed deep blue phosphorescent devices | D. Sun et al. (2014) |

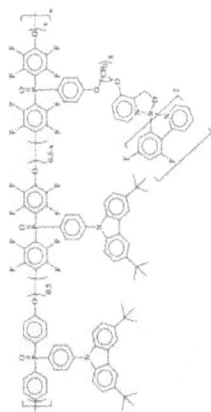

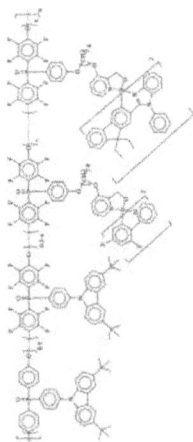

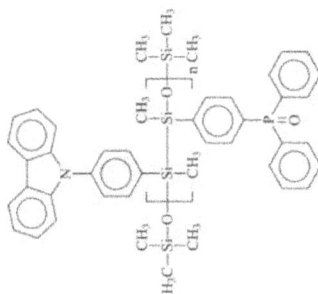

*(Continued)*

**TABLE 4.2**

(Continued)

| Polymeric backbone | Chemical structure | Features | Refs. |
|---|---|---|---|
| Poly(arylene ether-1,3,4-oxadiazole) | 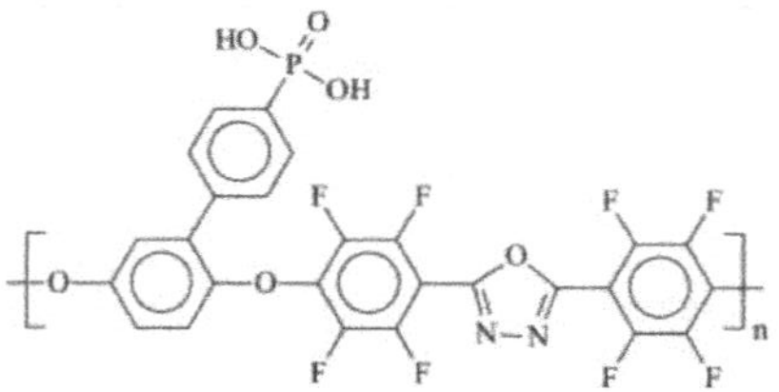 | Decent dielectric property | Ganesh et al. (2014) |
| Polyimide | 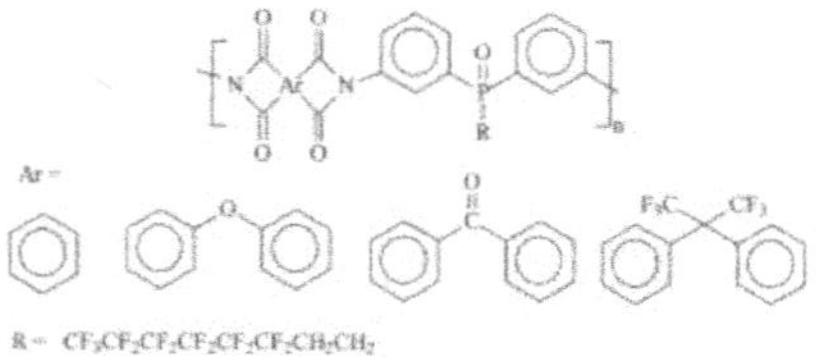 | Suitable for optical devices | Çakmakçı et al. (2016) |
| Polyimide | 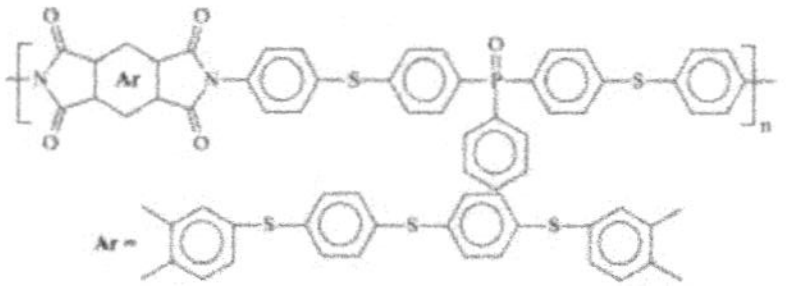 | High refractive index (1.725) polymer with low birefringence (0.0087) | Macdonald et al. (2017) |
| Crosslinked polymers through sol–gel process | 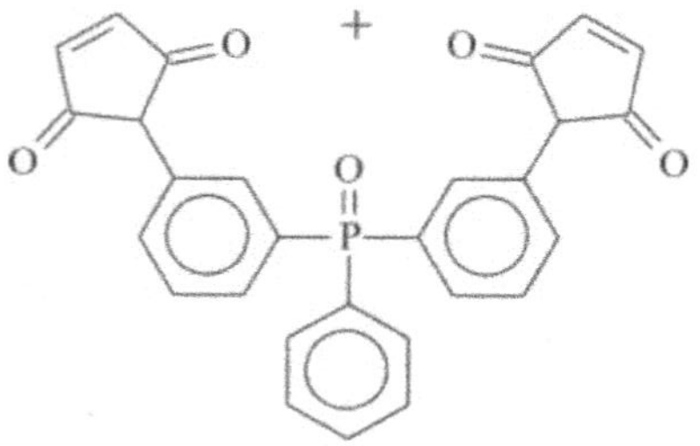 | Excellent temporal stability (100°C) and low optical loss (0.99–1.71 dB/cm; 830 nm) | C.-P. Chen et al. (2004) |

It should be stated that due to their great ability to retain water within the membranes, creation of well-networked ionic channels for proton conduction, and adhesion properties, polar phosphine oxide moiety-based PEMs are particularly intriguing in the PEMFC sector (C. Zhang et al., 2009). More importantly, the phosphine oxide moiety's passivation effect and hydrogen peroxide ($H_2O_2$) decomposition have enabled polymer membranes to achieve exceptional oxidative stability in PEM applications (Ghorai & Banerjee, 2023; Tsuneda et al., 2018).

### 4.2.4 Carboxyl Group-based Functionalization of Nanomaterials and Polymers

Carboxyl groups comprise two functional groups—hydroxyl (single-bonded OH) and carbonyl (double-bonded O) groups—connected to a single carbon atom. The carboxyl (COOH) group is so named because of the carbonyl group (C=O) and a hydroxyl group. They are made up of carboxylic acids and amino acids. They are marked with the IUPAC suffix acid.

The link between the carbon and the extremely electronegative oxygen in the carboxylic acid group is polar. A molecule with a carboxyl group will have a high melting point, centers that are hydrophilic, and a high boiling point.

The creation of a hydrogen bond between the solid and liquid states can dissipate the source of a material's high melting and boiling points. Fatty acids are one of the common examples. Hydrogen bonding makes earlier carboxylic acid members liquid in water. Due to the carbon chain's increasing hydrophobicity, higher members are less soluble.

A frequent functional group in organic chemistry is the carboxyl. It is also known as a carboxyl radical, a carboxyl functional group, or a carboxy group. A carboxyl group is made up of a carbon atom that is singly connected to a hydroxyl group and double bonded to oxygen. Hence, R-COOH is the formula for a carboxyl group, where R is the chain of the organic compound. Figure 4.3 depicts the modification of MWCNTs with a carboxyl group and $TiO_2$ nanoparticles.

Carboxylic acid would be the most typical example of a molecule that contains a carboxyl group. These compounds are widely present in nature. These are a few simple examples.

Amino acids are another type of molecules that has a carboxyl group. Because they have both an amino group and a carboxyl group, amino acids are special. An amino acid with a carboxyl functional group is glycine.

These amino acids each have a carboxyl group and help our bodies produce proteins. It is crucial to remember that much larger molecules in nature or our bodies often contain a carboxyl functional group. An innovative method of analyzing titrimetric data was used, assuming that rather than intermolecular changes in the pKa values of simple carboxylic acids, the variability in carboxyl group pKa values in fulvic acid was predominantly an intramolecular occurrence in polyprotic acid structures.

Carboxylic acids are polar acids. Since they are both hydrogen-bond acceptors (the carbonyl -C=O) and donors (the hydroxyl -OH), they additionally take part in hydrogen bonding. The hydroxyl and carbonyl groups make up the carboxyl functional

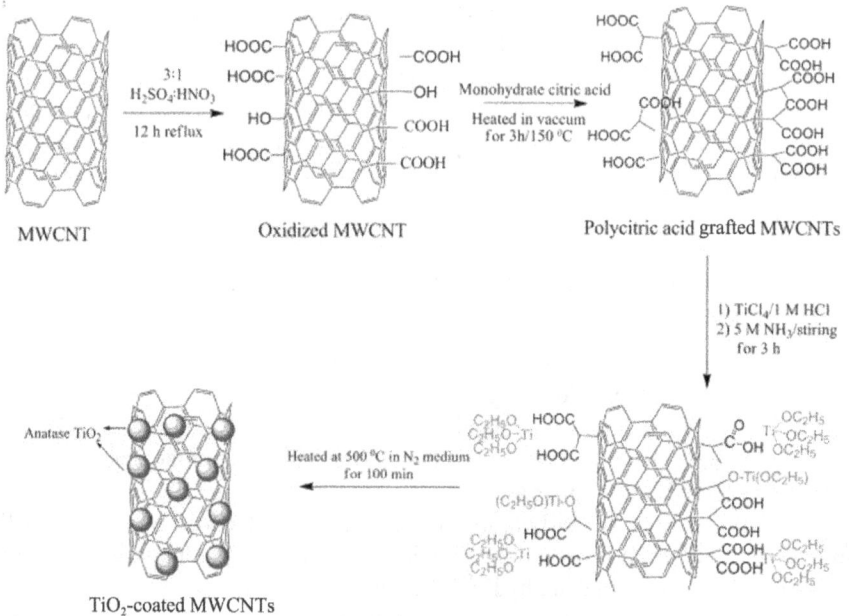

**FIGURE 4.3**

Carboxyl group and TiO$_2$-modified multiwalled carbon nanotubes (MWCNTs) (Vatanpour et al., 2012). *Reproduced with permission from Elsevier.*

group. Carboxylic acids generally exist in nonpolar environments as dimers due to their propensity to "self-associate." In contrast to smaller carboxylic acids (1–5 carbons), which are soluble in water, larger carboxylic acids have restricted solubility due to the increasing hydrophobicity of the alkyl chain. These longer chain acids frequently dissolve in ethers and alcohols, which are less polar solvents. Even hydrophobic carboxylic acids react with aqueous sodium hydroxide to produce sodium ions that are water soluble (Kralj et al., 2011; Mondal & Hu, 2006).

The conventional method for obtaining GO is chemical exfoliation of graphite oxide (GtO). The Brodie, Staudenmaier (1898), or Hummers Jr and Offeman (2002) method, or various modifications of these methods (e.g., the modified Hummer's method) (Li et al., 2010), allows for significant amounts of epoxy, hydroxyl, and carboxylic groups in the resulting graphene (Z. Sun et al., 2011). The highly oxidized graphene synthesized using this technique is commonly referred to as GO (Dreyer et al., 2014). Hummers oxidized graphite with a solution of potassium permanganate (KMnO$_4$) and sulfuric acid (H$_2$SO$_4$), whereas Brodie and Staudenmaier oxidized graphite with a solution of nitric acid (HNO$_3$) and potassium chlorate (KClO$_3$). The obtained graphite salts are produced by intercalating graphite with strong acids such as H$_2$SO$_4$, HNO$_3$, or HClO$_4$; simultaneously, graphite is oxidized and then exfoliated into GO nanosheets in water (Boehm et al., 1967; Hegab & Zou, 2015).

### 4.2.5 Hydroxyl Group-based Functionalization of Nanomaterials and Polymers

The hydroxyl functional group is the simplest of all the common organic functional groups. It consists of a hydrogen atom bonded to an oxygen atom. It has two chemical properties: acidity, which is defined by the pH value, and nucleophilicity, which refers to how receptive it is to other molecules.

These are chemical groups that are found on the carbon atoms that are joined to oxygen in organic molecules. It can form hydrogen bonds because it has two lone pairs of electrons. Common examples of compounds containing a hydroxyl group include water and ethanol. Chemical functional groups for hydroxyl include a hydroxyl group (OH) and a hydroxyl functional group (H-OH). [H-OH] is a compound made of hydrogen (H) and oxygen (O). It is present in substances with one or more hydroxyl groups. These substances have the general formula H-OH and its derivatives.

The amine group, hydroxyl group, carboxylic acid group, carbonyl group, aldehydes, ketones, and amino group or amino acids are the seven functional groups. There is an atom connected to each of the four different groups—the carbonyl group, three more organic groups, and the C=O double bond. The structure of the carboxylic acid is R-COOH. R stands for either an organic group or an atom of hydrogen. The hydrogen atom is replaced with a $CH_3$ group because no oxygen is present. Aldehydes are organic compounds that have an alkyne and a carbonyl double bond (a C-C triple bond). The -OH group is the hydroxyl functional group. This group belongs to the family of polar uncharged organic groups. Due to their polarity, these oxygen-containing compounds work well in aqueous solvents like water and many oils. Since these -OH groups are weak acids, whether in solution or as a component of a molecule, they can react with nonmetals like carbon.

Most chemical interactions involving other molecules or atoms require the reactive hydroxyl group, a chemical functionality. Because of its role in facilitating solubility and structure, hydrogen bonding is crucial. One of the most important factors affecting the solubility of one substance in another is the presence of hydrogen bonds in the two molecules. High melting and boiling temperatures are also a result of hydrogen bonding of compounds that contain it. This group's general structure is R-O-(H-OH). Alcohol, phenols, amines, carboxylic acids, esters, ethers, anhydrides, and other derivatives are a few examples.

The hydroxyl group exhibits nucleophilicity, electrophilicity, and polarity. Since water has a strong affinity for this group, its polarity makes it an ideal donor of hydrogen bonds. Many ions or salts, including sodium hydroxide (NaOH), mercury(II) chloride ($HgCl_2$), and calcium hypochlorite ($Ca(OCl)_2$), are also donors of hydrogen bonds (Figure 4.4).

One hydrogen atom and one oxygen atom make up a hydroxyl group. Organic compounds frequently have the hydroxyl group attached, turning them into significant chemicals. A substance is referred to as organic if it has carbon atoms covalently linked to hydrogen atoms. All organic compounds have carbon as their main component.

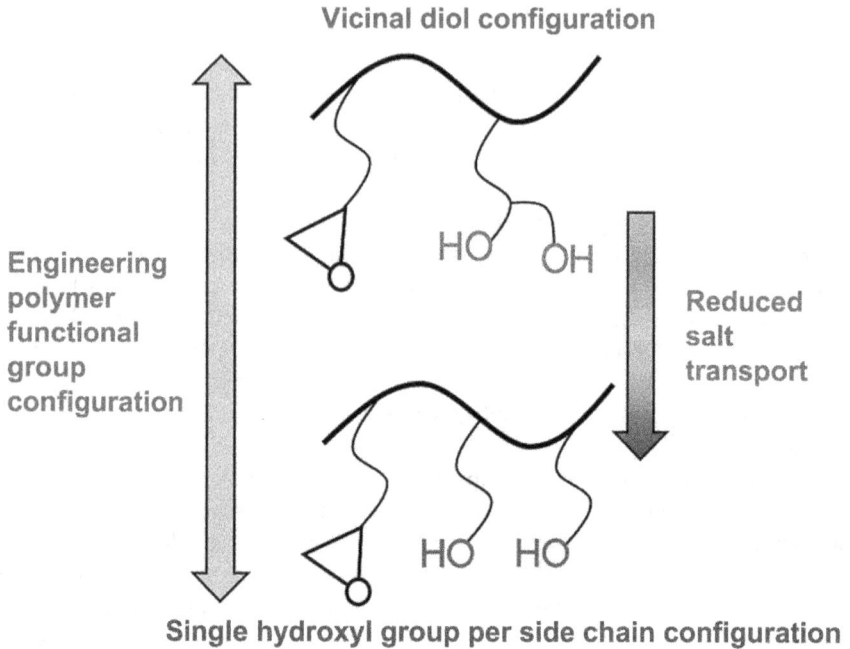

**FIGURE 4.4**
Influence of hydroxyl group in the desalination process (Luo et al., 2019). *Reproduced with permission from Elsevier.*

Any class of chemical compounds that include one or more hydroxyl groups attached to the carbon atom of an alkyl group is considered to be alcohol. Blue circles surround the hydroxyl (-OH). A straight chain of carbon atoms single-bonded to hydrogen atoms is known as an alkane. An alkyl group is created when hydrogen is removed from an alkane chain.

Alcohols are organic compounds in which the hydrogen atom of an aliphatic carbon has been replaced by a hydroxyl group. An alcohol molecule is therefore composed of two parts: an alkyl group and a hydroxyl group. They have a pleasant odor. They exhibit a distinctive collection of chemical and physical properties. The physical and chemical characteristics of alcohols are mostly a result of the presence of the hydroxyl group. The alcohol structure is influenced by a few variables.

Alcohol is acidic by nature and reacts with metals like sodium, potassium, and others. The reaction is due to the polarity of the bond connecting the hydrogen and oxygen atoms in the hydroxyl group. Primary alcohols are more acidic than secondary and tertiary alcohols (You et al., 2010).

## 4.2.6  Sulfonic Group-based Functionalization of Nanomaterials and Polymers

Polymer sulfonation approaches are mainly derived from those used for small organic compounds. So both techniques have relatively similar reaction conditions, but large, bulky materials like polymers typically have fewer solvent selections than small organic molecules. The kind of polymer and the desired amount of sulfonation (DS) often define the appropriate reaction conditions. Reactions may occur at temperatures between 20 °C and 300 °C; however, greater yields and degrees of sulfonation necessitate a greater temperature. However, the maximal reaction temperature is restricted to 300 °C due to the sulfonated polymer's breakdown via sulfone production, which increases with rising temperature (Shibuya & Porter, 1994). The process is influenced by temperature in addition to sulfonation agent concentration. In addition, increasing sulfonation agent concentrations to increase reaction rates and DS can also result in polymer chain cleavage or crosslinking (Miyatake et al., 2007). Since desulfonation processes could take place as a result of temperature-induced hydrolysis, water as a by-product must also be considered (Kučera & Jančář, 1998).

Sulfonated polymers can be synthesized using various kinds of processes, including (i) lithiation–sulfonation–oxidation, (ii) direct sulfonation, (iii) polymerization of sulfonated monomers, and (iv) grafting sulfonated molecules onto a polymer chain (Pourcelly, 2011).

Sulfonated aromatic (hydrocarbon) polymers (SAPs) have been used to develop proton-conductive membranes because of their superior chemical and mechanical stability, which is needed for proton-exchange membranes (S. et al., 2020). Examples of SAPs include sulfonated poly(2,6-dimethyl-1,4-phenylene oxide) (SPPO) (Kruczek & Matsuura, 1998), polyether sulfone (SPES) (Tripathi et al., 2010), polyphenyl sulfone (SPPSU) (Matsushita & Kim, 2018), and sulfonated poly (ether ether ketone) (SPEEK) (W. Zhang et al., 2006). These SAPs are made from economical but robust aromatic polymers. The capacity of a polymer to form films and its good

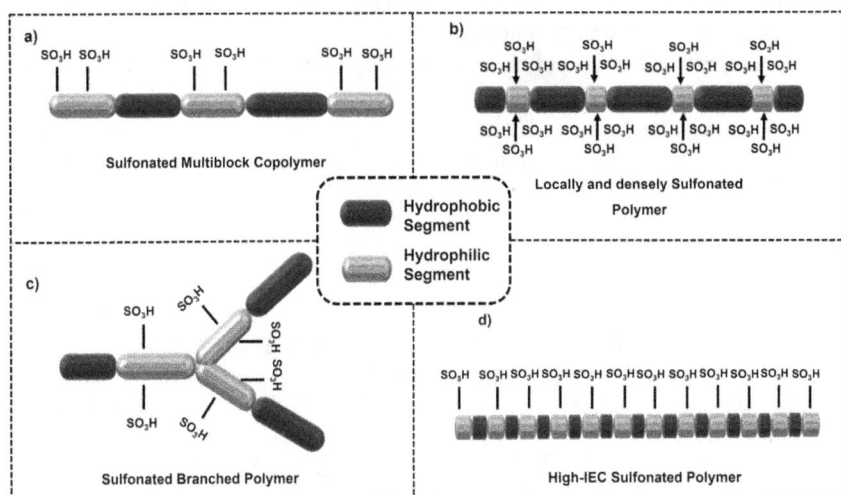

**FIGURE 4.5**
Sulfonation of a polymer and its structural formation (Higashihara et al., 2009). *Reproduced with permission from Elsevier.*

mechanical and chemical stability are due to the aromaticity of the macromolecular chains, which neatly stack together and provide strong intramolecular linkages (Figure 4.5) (Z. Zhang et al., 2011). However, the stability of aromatic polymers alone is insufficient for use in PEMFCs. Polymers must also have adequate proton conductivity.

Hydrophobic aromatic polymer chains undergo sulfonation, which is the addition of sulfonic acid moieties. The proton conductivity needed in PEMFCs can be induced by these sulfonic acid moieties. The sulfonation process is not always simple because an increase in the average amount of sulfonic acid groups per repeating unit implies a deterioration in the polymer's mechanical characteristics, frequently producing polymers that are gel-like when wet and brittle when dry (Matsumoto et al., 2009). SAPs are often formed into films and crosslinked to increase mechanical properties. SAP crosslinking covalently binds the macromolecular chains, improving mechanical and chemical stability as well as gas permeability (Khomein et al., 2021; Mabrouk et al., 2014). An exceptionally potent organic acid with acidity similar to that of inorganic acids is sulfonic acid. Sulfonic acids are sulfuric acids with an organic substituent in place of one of the hydroxyl groups. Sulfuric acid's tautomer, $S(=O)(OH)_2$, and parent sulfonic acid, $HS(=O)_2(OH)$ (with the organic group replaced by hydrogen), are the parent chemicals in this case.

The salts and different derivatives are used to create detergents, water-soluble dyes and catalysts, sulfonamide medications, and ion-exchange resins, while the free acids are widely utilized as catalysts in organic syntheses. As starting materials or intermediates in synthesis, aromatic sulfonic acids are extremely valuable. Sulfonic acid hydrolysis can result in the production of benzene. In this procedure, benzene is produced by exposing benzene sulfonic acid to extremely hot steam. An aqueous solution of NaOH is used to treat benzene sulfonic acid. The solid NaOH is then added to this mixture, and it is then heated to a high temperature. The resulting mixture is then acidified to produce phenol.

High-temperature stable polymers called polyaromatic membrane materials with sulfonated aromatic units are being researched as solid polymer electrolyte (SPE) candidates. Such membrane composition is essential to obtain optimal ionic conductivity since the hydrophilic sulfonic acid group ($-SO_3H$) causes proton conductivity.

---

## References

Abid, O. H., Tawfeeq, H. M., & Muslim, R. F. (2017). Synthesis and characterization of novel 1,3-oxazepin-5(1h)-one derivatives via reaction of imine compounds with isobenzofuran-1(3h)-one. *ACTA Pharmaceutica Sciencia, 55*(4), 43. https://doi.org/10.23893/1307-2080.APS.05525

Allcock, H. R., & Kugel, R. L. (2002, May 1). *Synthesis of high polymeric alkoxy- and aryloxyphosphonitriles* (world). ACS Publications; American Chemical Society. https://doi.org/10.1021/ja01096a056

Bae, B., Miyatake, K., & Watanabe, M. (2010). Sulfonated poly(arylene ether sulfone ketone) multiblock copolymers with highly sulfonated block. Synthesis and properties. *Macromolecules, 43*(6), 2684–2691. https://doi.org/10.1021/ma100291z

Bian, X.-C., Chen, L., Wang, J.-S., & Wang, Y.-Z. (2010). A novel thermotropic liquid crystalline copolyester containing phosphorus and aromatic ether moiety toward high flame retardancy and low mesophase temperature. *Journal of Polymer Science Part A: Polymer Chemistry, 48*(5), 1182–1189. https://doi.org/10.1002/pola.23878

Bisoi, S., Mandal, A. K., Singh, A., Padmanabhan, V., & Banerjee, S. (2017). Soluble, optically transparent polyamides with a phosphaphenanthrene skeleton: Synthesis, characterization, gas permeation and molecular dynamics simulations. *Polymer Chemistry, 8*(29), 4220–4232. https://doi.org/10.1039/C7PY00687J

Boehm, H. P., Eckel, M., & Scholz, W. (1967). Untersuchungen am Graphitoxid V. Über den Bildungsmechanismus des Graphitoxids. *Zeitschrift für anorganische und allgemeine Chemie, 353*(5–6), 236–242. https://doi.org/10.1002/zaac.19673530503

Çakmakçı, E., Güngör, A., & Gören, A. C. (2016). Fluorine and phosphine oxide containing homo and copolyimides. *Journal of Fluorine Chemistry, 186*, 66–78. https://doi.org/10.1016/j.jfluchem.2016.03.016

Chang, T. C., Shen, W. S., Chiu, Y. S., & Ho, S. Y. (1995). Thermo-oxidative degradation of phosphorus-containing polyurethane. *Polymer Degradation and Stability, 49*(3), 353–360. https://doi.org/10.1016/0141-3910(95)00116-4

Chen, B., Long, F., Chen, S., Cao, Y., & Pan, X. (2020). Magnetic chitosan biopolymer as a versatile adsorbent for simultaneous and synergistic removal of different sorts of dyestuffs from simulated wastewater. *Chemical Engineering Journal, 385*, 123926. https://doi.org/10.1016/j.cej.2019.123926

Chen, C.-P., Huang, G.-S., Jeng, R.-J., Chou, C.-C., Su, W.-C., & Chang, H.-L. (2004). Low loss second-order non-linear optical crosslinked polymers based on a phosphorus-containing maleimide. *Polymers for Advanced Technologies, 15*(10), 587–592. https://doi.org/10.1002/pat.516

Choi, J. K., Paek, K. Y., & Yoon, T.-H. (2009). Adhesive and dielectric properties of novel polyimides with bis(3,3′-aminophenyl)-2,3,5,6-tetrafluoro-4-trifluoromethyl phenyl phosphine oxide (mDA7FPPO). *European Polymer Journal, 45*(6), 1652–1658. https://doi.org/10.1016/j.eurpolymj.2009.03.009

Choudhury, L. H., & Parvin, T. (2011). Recent advances in the chemistry of imine-based multicomponent reactions (MCRs). *Tetrahedron, 67*(43), 8213–8228. https://doi.org/10.1016/j.tet.2011.07.020

Dong, Q., Lei, J., Wang, H., Ke, M., Liang, X., Yang, X., Liang, H., Huselstein, C., Tong, Z., & Chen, Y. (2022). Antibacterial soy protein isolate prepared by quaternization. *International Journal of Molecular Sciences, 23*(16), Article 16. https://doi.org/10.3390/ijms23169110

Dreyer, D. R., Todd, A. D., & Bielawski, C. W. (2014). Harnessing the chemistry of graphene oxide. *Chemical Society Reviews, 43*(15), 5288–5301. https://doi.org/10.1039/C4CS00060A

Elabd, Y. A., & Hickner, M. A. (2011). Block copolymers for fuel cells. *Macromolecules, 44*(1), 1–11. https://doi.org/10.1021/ma101247c

Elhalwagy, M. E., Elsherbiny, A. S., & Gemeay, A. H. (2023). Amine-rich polymers for water purification applications. *Materials Today Chemistry, 27*, 101344. https://doi.org/10.1016/j.mtchem.2022.101344

Fu, L., Xiao, G., & Yan, D. (2010). Sulfonated poly(arylene ether sulfone)s with phosphine oxide moieties: A promising material for proton exchange membranes. *ACS Applied Materials & Interfaces, 2*(6), 1601–1607. https://doi.org/10.1021/am1000739

Ganesh, S. D., Pai, V. K., Kariduraganavar, M. Y., & Jayanna, M. B. (2014). Fluorinated poly(arylene ether-1,3,4-oxadiazole)s containing a 4-bromophenyl pendant group and its phosphonated derivatives: Synthesis, spectroscopic characterization, thermal and dielectric studies. *Polymer-Plastics Technology and Engineering, 53*(1), 97–105. https://doi.org/10.1080/03602559.2013.843694

Ghorai, A., & Banerjee, S. (2023). Phosphorus-containing aromatic polymers: Synthesis, structure, properties and membrane-based applications. *Progress in Polymer Science, 138*, 101646. https://doi.org/10.1016/j.progpolymsci.2023.101646

Ghorai, A., Mandal, A. K., & Banerjee, S. (2020). Synthesis and characterization of new phosphorus containing sulfonated polytriazoles for proton exchange membrane application. *Journal of Polymer Science, 58*(2), 263–279. https://doi.org/10.1002/pol.20190030

Hegab, H. M., & Zou, L. (2015). Graphene oxide-assisted membranes: Fabrication and potential applications in desalination and water purification. *Journal of Membrane Science, 484*, 95–106. https://doi.org/10.1016/j.memsci.2015.03.011

Hickner, M. A., Ghassemi, H., Kim, Y. S., Einsla, B. R., & McGrath, J. E. (2004). Alternative polymer systems for proton exchange membranes (PEMs). *Chemical Reviews, 104*(10), 4587–4612. https://doi.org/10.1021/cr020711a

Higashihara, T., Matsumoto, K., & Ueda, M. (2009). Sulfonated aromatic hydrocarbon polymers as proton exchange membranes for fuel cells. *Polymer, 50*(23), 5341–5357. https://doi.org/10.1016/j.polymer.2009.09.001

Hummers Jr, W. S., & Offeman, R. E. (2002, May 1). *Preparation of graphitic oxide* (world). ACS Publications; American Chemical Society. https://doi.org/10.1021/ja01539a017

Jespersen, J. L., Schaltz, E., & Kær, S. K. (2009). Electrochemical characterization of a polybenzimidazole-based high temperature proton exchange membrane unit cell. *Journal of Power Sources, 191*(2), 289–296. https://doi.org/10.1016/j.jpowsour.2009.02.025

Khomein, P., Ketelaars, W., Lap, T., & Liu, G. (2021). Sulfonated aromatic polymer as a future proton exchange membrane: A review of sulfonation and crosslinking methods. *Renewable and Sustainable Energy Reviews, 137*, 110471. https://doi.org/10.1016/j.rser.2020.110471

Kim, J.-P., Kang, J.-W., Kim, J.-J., & Lee, J.-S. (2003). Synthesis and optical properties of fluorinated poly(arylene ether phosphine oxide)s. *Journal of Polymer Science Part A: Polymer Chemistry, 41*(10), 1497–1503. https://doi.org/10.1002/pola.10694

Kralj, S., Drofenik, M., & Makovec, D. (2011). Controlled surface functionalization of silica-coated magnetic nanoparticles with terminal amino and carboxyl groups. *Journal of Nanoparticle Research, 13*(7), 2829–2841. https://doi.org/10.1007/s11051-010-0171-4

Kruczek, B., & Matsuura, T. (1998). Development and characterization of homogeneous membranes de from high molecular weight sulfonated polyphenylene oxide. *Journal of Membrane Science, 146*(2), 263–275. https://doi.org/10.1016/S0376-7388(98)00120-3

Kučera, F., & Jančář, J. (1998). Homogeneous and heterogeneous sulfonation of polymers: A review. *Polymer Engineering & Science, 38*(5), 783–792. https://doi.org/10.1002/pen.10244

Li, H., Zou, L., Pan, L., & Sun, Z. (2010). Novel graphene-like electrodes for capacitive deionization. *Environmental Science & Technology, 44*(22), 8692–8697. https://doi.org/10.1021/es101888j

Liu, Y., Gan, Y., Zhao, C., Xie, M., Gu, D., Shuai, S., Zhu, H., & Hao, J. (2022). Polyethyleneimine-modified magnetite prepared without a base to remove Congo red in water. *Surface Innovations, 10*(1), 76–85. https://doi.org/10.1680/jsuin.21.00006a

Luo, H., Chang, K., Bahati, K., & Geise, G. M. (2019). Functional group configuration influences salt transport in desalination membrane materials. *Journal of Membrane Science, 590*, 117295. https://doi.org/10.1016/j.memsci.2019.117295

Mabrouk, W., Ogier, L., Vidal, S., Sollogoub, C., Matoussi, F., & Fauvarque, J. F. (2014). Ion exchange membranes based upon crosslinked sulfonated polyethersulfone for electrochemical applications. *Journal of Membrane Science, 452*, 263–270. https://doi.org/10.1016/j.memsci.2013.10.006

Macdonald, E. K., Lacey, J. C., Ogura, I., & Shaver, M. P. (2017). Aromatic polyphosphonates as high refractive index polymers. *European Polymer Journal, 87*, 14–23. https://doi.org/10.1016/j.eurpolymj.2016.12.003

Makvandi, P., Iftekhar, S., Pizzetti, F., Zarepour, A., Zare, E. N., Ashrafizadeh, M., Agarwal, T., Padil, V. V. T., Mohammadinejad, R., Sillanpaa, M., Maiti, T. K., Perale, G., Zarrabi, A., & Rossi, F. (2021). Functionalization of polymers and nanomaterials for water treatment, food packaging, textile and biomedical applications: A review. *Environmental Chemistry Letters, 19*(1), 583–611. https://doi.org/10.1007/s10311-020-01089-4

Mandal, A. K., Bisoi, S., & Banerjee, S. (2019). Effect of phosphaphenanthrene skeleton in sulfonated polyimides for proton exchange membrane application. *ACS Applied Polymer Materials, 1*(4), 893–905. https://doi.org/10.1021/acsapm.9b00128

Matsumoto, K., Nakagawa, T., Higashihara, T., & Ueda, M. (2009). Sulfonated poly(ether sulfone)s with binaphthyl units as proton exchange membranes for fuel cell application. *Journal of Polymer Science Part A: Polymer Chemistry, 47*(21), 5827–5834. https://doi.org/10.1002/pola.23627

Matsushita, S., & Kim, J.-D. (2018). Organic solvent-free preparation of electrolyte membranes with high proton conductivity using aromatic hydrocarbon polymers and small cross-linker molecules. *Solid State Ionics, 316*, 102–109. https://doi.org/10.1016/j.ssi.2017.12.033

Miyake, J., Watanabe, M., & Miyatake, K. (2013). Sulfonated poly(arylene ether phosphine oxide ketone) block copolymers as oxidatively stable proton conductive membranes. *ACS Applied Materials & Interfaces, 5*(13), 5903–5907. https://doi.org/10.1021/am401625j

Miyatake, K., Chikashige, Y., Higuchi, E., & Watanabe, M. (2007). Tuned polymer electrolyte membranes based on aromatic polyethers for fuel cell applications. *Journal of the American Chemical Society, 129*(13), 3879–3887. https://doi.org/10.1021/ja0672526

Mondal, S., & Hu, J. L. (2006). Structural characterization and mass transfer properties of nonporous segmented polyurethane membrane: Influence of hydrophilic and carboxylic group. *Journal of Membrane Science, 274*(1), 219–226. https://doi.org/10.1016/j.memsci.2005.08.016

Monge, S., Canniccioni, B., Graillot, A., & Robin, J.-J. (2011). Phosphorus-containing polymers: A great opportunity for the biomedical field. *Biomacromolecules, 12*(6), 1973–1982. https://doi.org/10.1021/bm2004803

Muhammad, A., Shah, A.-H. A., & Bilal, S. (2019). Comparative study of the adsorption of acid blue 40 on polyaniline, magnetic oxide and their composites: Synthesis, characterization and application. *Materials, 12*(18), Article 18. https://doi.org/10.3390/ma12182854

Muhammad, A., Shah, A.-H. A., Bilal, S., & Rahman, G. (2019). Basic blue dye adsorption from water using polyaniline/magnetite (Fe3O4) composites: Kinetic and thermodynamic aspects. *Materials, 12*(11), Article 11. https://doi.org/10.3390/ma12111764

Mukherjee, S., Akshay, & Samanta, A. N. (2019). Amine-impregnated MCM-41 in postcombustion CO2 capture: Synthesis, characterization, isotherm modelling. *Advanced Powder Technology, 30*(12), 3231–3240. https://doi.org/10.1016/j.apt.2019.09.032

Petrov, K. A., Nifant'ev, E. Ye., Lysenko, T. N., & Suzanskii, A. I. (1963). Phosphorus-containing polymers—VI. Synthesis of polyphosphites and polyphosphonites based on glucose. *Polymer Science U.S.S.R.*, *4*(6), 1417–1424. https://doi.org/10.1016/0032-3950(63) 90016-9

Pourcelly, G. (2011). Membranes for low and medium temperature fuel cells. State-of-the-art and new trends. *Petroleum Chemistry*, *51*(7), 480–491. https://doi.org/10.1134/ S0965544111070103

Raja Rafidah, S. R., Rashmi, W., Khalid, M., Wong, Y. W., & Priyanka, J. (2020). Recent progress in the development of aromatic polymer-based proton exchange membranes for fuel cell applications. *Polymers*, *12*(5), Article 5. https://doi.org/10.3390/polym12051061

Salim, S. R. S. (2021). Treatment of amine wastes generated in industrial processes. *IOP Conference Series: Materials Science and Engineering*, *1092*(1), 012051. https://doi. org/10.1088/1757-899X/1092/1/012051

Salmeia, K. A., & Gaan, S. (2015). An overview of some recent advances in DOPO-derivatives: Chemistry and flame retardant applications. *Polymer Degradation and Stability*, *113*, 119–134. https://doi.org/10.1016/j.polymdegradstab.2014.12.014

Sato, M., Tada, Y., & Yokoyama, M. (1980). Preparation of phosphorus-containing polymers—XXIII: Phenoxaphosphine-containing polyimides. *European Polymer Journal*, *16*(8), 671–676. https://doi.org/10.1016/0014-3057(80)90032-4

Shan Wang, C., & Yueh Shieh, J. (1999). Synthesis and flame retardancy of phosphorus containing polycarbonate. *Journal of Polymer Research*, *6*(3), 149–154. https://doi. org/10.1007/s10965-006-0082-3

Shao, S., Ding, J., Wang, L., Jing, X., & Wang, F. (2012a). Highly efficient blue electrophosphorescent polymers with fluorinated poly(arylene ether phosphine oxide) as backbone. *Journal of the American Chemical Society*, *134*(37), 15189–15192. https://doi. org/10.1021/ja305634j

Shao, S., Ding, J., Wang, L., Jing, X., & Wang, F. (2012b). White electroluminescence from all-phosphorescent single polymers on a fluorinated poly(arylene ether phosphine oxide) backbone simultaneously grafted with blue and yellow phosphors. *Journal of the American Chemical Society*, *134*(50), 20290–20293. https://doi.org/10.1021/ja310158j

Shao, S., Ding, J., Ye, T., Xie, Z., Wang, L., Jing, X., & Wang, F. (2011). A novel, bipolar polymeric host for highly efficient blue electrophosphorescence: A non-conjugated poly(aryl ether) containing triphenylphosphine oxide units in the electron-transporting main chain and carbazole units in hole-transporting side chains. *Advanced Materials*, *23*(31), 3570–3574. https://doi.org/10.1002/adma.201101074

Shibuya, N., & Porter, R. S. (1994). A kinetic study of PEEK sulfonation in concentrated sulfuric acid by ultraviolet-visible spectroscopy. *Polymer*, *35*(15), 3237–3242. https://doi. org/10.1016/0032-3861(94)90128-7

Shobha, H. K., Johnson, H., Sankarapandian, M., Kim, Y. S., Rangarajan, P., Baird, D. G., & McGrath, J. E. (2001). Synthesis of high refractive-index melt-stable aromatic polyphosphonates. *Journal of Polymer Science Part A: Polymer Chemistry*, *39*(17), 2904–2910. https://doi.org/10.1002/pola.1270

Smith, C. D., Grubbs, H., Webster, H. F., Gungör, A., Wightman, J. P., & McGrath, J. E. (1991). Unique characteristics derived from poly(arylene ether phosphine oxide)s. *High Performance Polymers*, *3*(4), 211–229. https://doi.org/10.1088/0954-0083/3/4/001

Stackman, R. W. (2002, May 1). *Phosphorus based additives for flame retardant polyester. 2. Polymeric phosphorus esters* (world). ACS Publications; American Chemical Society. https://doi.org/10.1021/i300006a028

Staudenmaier, L. (1898). Verfahren zur Darstellung der Graphitsäure. *Berichte der deutschen chemischen Gesellschaft, 31*(2), 1481–1487. https://doi.org/10.1002/cber.18980310237

Sun, D., Fu, Q., Ren, Z., Li, H., Ma, D., & Yan, S. (2014). Synthesis of well-defined poly(phenylcarbazole-alt-triphenylphosphine oxide) siloxane as a bipolar host material for solution-processed deep blue phosphorescent devices. *Polymer Chemistry, 5*(1), 220–226. https://doi.org/10.1039/C3PY00840A

Sun, Z., James, D. K., & Tour, J. M. (2011). Graphene chemistry: Synthesis and manipulation. *The Journal of Physical Chemistry Letters, 2*(19), 2425–2432. https://doi.org/10.1021/jz201000a

Tripathi, B. P., Chakrabarty, T., & Shahi, V. K. (2010). Highly charged and stable cross-linked 4,4′-bis(4-aminophenoxy)biphenyl-3,3′-disulfonic acid (BAPBDS)-sulfonated poly(ether sulfone) polymer electrolyte membranes impervious to methanol. *Journal of Materials Chemistry, 20*(37), 8036–8044. https://doi.org/10.1039/C0JM01183E

Tsuneda, T., Miyake, J., & Miyatake, K. (2018). Mechanism of $H_2O_2$ decomposition by triphenylphosphine oxide. *ACS Omega, 3*(1), 259–265. https://doi.org/10.1021/acsomega.7b01416

Umezawa, K., Oshima, T., Yoshizawa-Fujita, M., Takeoka, Y., & Rikukawa, M. (2012). Synthesis of hydrophilic–hydrophobic block copolymer ionomers based on polyphenylenes. *ACS Macro Letters, 1*(8), 969–972. https://doi.org/10.1021/mz300290x

Vatanpour, V., Madaeni, S. S., Moradian, R., Zinadini, S., & Astinchap, B. (2012). Novel antibifouling nanofiltration polyethersulfone membrane fabricated from embedding $TiO_2$ coated multiwalled carbon nanotubes. *Separation and Purification Technology, 90*, 69–82. https://doi.org/10.1016/j.seppur.2012.02.014

Vidal, F., & Jäkle, F. (2019). Functional polymeric materials based on main-group elements. *Angewandte Chemie International Edition, 58*(18), 5846–5870. https://doi.org/10.1002/anie.201810611

Wan, T., Jia, Y., He, S., Wang, T., Wang, J., Tang, Q., & Yu, M. (2022). Enhanced adsorption of crystal violet from aqueous solution by polyethyleneimine-modified magnetic hydrogel nanocomposites. *Polymer Bulletin*. https://doi.org/10.1007/s00289-022-04422-9

Wehbi, M., Mehdi, A., Negrell, C., David, G., Alaaeddine, A., & Améduri, B. (2020). Phosphorus-containing fluoropolymers: State of the art and applications. *ACS Applied Materials & Interfaces, 12*(1), 38–59. https://doi.org/10.1021/acsami.9b16228

You, Z., Cao, H., Gao, J., Shin, P. H., Day, B. W., & Wang, Y. (2010). A functionalizable polyester with free hydroxyl groups and tunable physiochemical and biological properties. *Biomaterials, 31*(12), 3129–3138. https://doi.org/10.1016/j.biomaterials.2010.01.023

Zhang, C., Kang, S., Ma, X., Xiao, G., & Yan, D. (2009). Synthesis and characterization of sulfonated poly(arylene ether phosphine oxide)s with fluorenyl groups by direct polymerization for proton exchange membranes. *Journal of Membrane Science, 329*(1), 99–105. https://doi.org/10.1016/j.memsci.2008.12.021

Zhang, M., Yu, Z., & Yu, H. (2020). Adsorption of eosin Y, methyl orange and brilliant green from aqueous solution using ferroferric oxide/polypyrrole magnetic composite. *Polymer Bulletin, 77*(2), 1049–1066. https://doi.org/10.1007/s00289-019-02792-1

Zhang, W., Gogel, V., Friedrich, K. A., & Kerres, J. (2006). Novel covalently cross-linked poly(etheretherketone) ionomer membranes. *Journal of Power Sources, 155*(1), 3–12. https://doi.org/10.1016/j.jpowsour.2004.12.082

Zhang, Y.-R., Su, P., Huang, J., Wang, Q.-R., & Zhao, B.-X. (2015). A magnetic nanomaterial modified with poly-lysine for efficient removal of anionic dyes from water. *Chemical Engineering Journal, 262*, 313–318. https://doi.org/10.1016/j.cej.2014.09.094

Zhang, Z., Wu, L., & Xu, T. (2011). Synthesis and properties of side-chain-type sulfonated poly(phenylene oxide) for proton exchange membranes. *Journal of Membrane Science, 373*(1), 160–166. https://doi.org/10.1016/j.memsci.2011.03.002

Zhou, Z., & Liu, R. (2017). $Fe_3O_4$@polydopamine and derived $Fe_3O_4$@carbon core–shell nanoparticles: Comparison in adsorption for cationic and anionic dyes. *Colloids and Surfaces A: Physicochemical and Engineering Aspects, 522*, 260–265. https://doi.org/10.1016/j.colsurfa.2017.02.063

# 5

# Functional Nanomaterials and Polymers for Wastewater Treatment

## 5.1 Overview of Wastewater Treatment and Its Approaches

Water contamination is a significant ecological disaster that has an influence on both ecosystem sustainability and people's health. The discharge of wastewater from both residences and industries is a crucial factor in this case. Pathogens, polychlorinated biphenyls, pesticides, heavy metal ions, and pharmaceutical waste are a few examples of contaminants that can be harmful to both the environment and human health. Effective wastewater treatment is crucial to reducing the negative impacts of wastewater on the environment and public health.

Conventional approaches have various limitations for eliminating pesticides, which led scientists to develop novel functionalized materials and subsequent technologies. Heterostructure functionalization enhances the chemical and physical characteristics of the materials. Since limited electron mobility and the development of discrete energy levels have improved material quality, photoactive materials function more effectively. Improved surface defects, a higher surface-to-volume ratio, good electron mobility, greater light absorption, and a high affinity for metabolite sorption in the aqueous form contribute to the significant characteristics of functional materials (Lin et al., 2006). To improve their performance, these materials can be readily modified into polymers, zeolites, and membranes. Among other things, the production of sensors, handling hazardous waste, and other processes depend on functional materials (Mohmood et al., 2013). According to studies (Al-Hamdi et al., 2016; Rasheed et al., 2021) on the removal of pesticides from aqueous media, various iron oxides including titanium dioxide ($TiO_2$), zinc oxide (ZnO), magnetite and hematite, tungsten oxides, and zirconium oxide (ZrO) work effectively.

Functional group polymers have gained recognition as potential materials for wastewater treatment due to their unique physicochemical characteristics and wide range of uses. These polymers contain reactive groups like cationic, anionic, or nonionic moieties that can interact with various contaminants present in wastewater (Figure 5.1). Functional group polymers function as coagulants, flocculants, adsorbents, or membrane modifiers and can be utilized alone or in combination with other therapeutic modalities.

Pesticides are chemical agents that are synthesized and used to kill or control pests. Organophosphates, organochlorines, carbamates, and pyrethroids are the

DOI: 10.1201/9781003391364-5

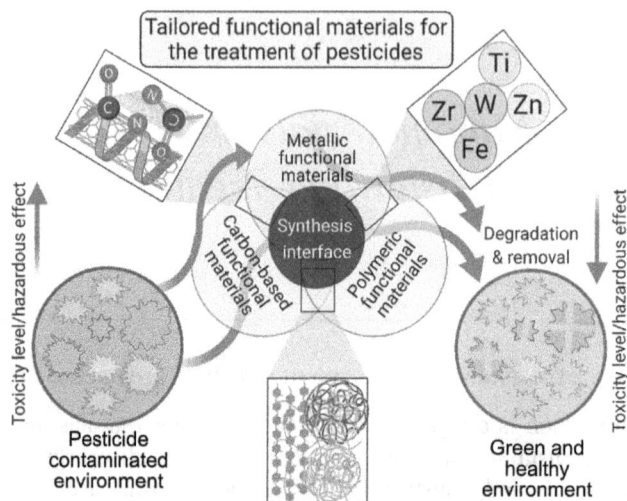

**FIGURE 5.1**
Various functional nanomaterials and polymers used for the removal of pesticides (Rasheed et al., 2021). *Reproduced with permission from Elsevier.*

most prevalent, and they are divided into groups based on the families to which they belong and the environments to which they are directed (Rana et al., 2021). To ensure there is enough food for everyone, the growing global population has made significant pesticide use in agriculture necessary. Two million tonnes of pesticides were used worldwide in 2019, with China and the United States being the most representative (Chen et al., 2022; Sanoja-López et al., 2023).

### 5.1.1 Background

A significant environmental concern in recent years has been the growing contamination of water resources brought on by the discharge of industrial effluents. To address this problem, various therapeutic methods have been devised, but the usage of functional group polymers has drawn the most interest. Specific chemical groups, including amino, carboxyl, sulfonic, and hydroxyl groups, are found in functional group polymers. These groups can interact with pollutants in water and successfully remove them.

In agriculture, animal husbandry, and public health, pesticides are used to eradicate or control undesired organisms or "pests" (Hassaan & El Nemr, 2020). Despite being widely used, only 0.1% of pesticides has an impact on the intended organisms, with the remaining portion scattered throughout other environmental areas, including water bodies (Bapat et al., 2022). Ingestion of contaminated food, inhalation of dust or aerosols, and direct skin contact when using this kind of product are all ways that pesticides might enter the human body. Overuse of pesticides poses a risk to human health because even minute concentrations of these compounds have both immediate and long-term side effects (Sanoja-López et al., 2023).

Therefore, preventing inappropriate human consumption depends on pesticide breakdown and removal. A number of strategies have been used to address this type of pollution, including physical approaches like adsorption, coagulation–flocculation, and membrane filtration and chemical approaches like electrocoagulation, electrooxidation, and photocatalysis (Bilici et al., 2021). Many different kinds of membranes on the market have been created using various materials, with ceramic and polymeric membranes being the most thoroughly researched (Arumugham et al., 2021; Pejman et al., 2021). More than 50 polymers are used in the production of membranes, with polyethersulfone (PES), polyacrylonitrile (PAN), polyvinylidene fluoride (PVDF), polyvinyl alcohol (PVA), and polystyrene (PS) being the most widely used ones (Lee et al., 2018; Sanoja-López et al., 2023). In comparison to conventional water treatment techniques, functional group polymers are more efficient, cheaper, and easier to use. These polymers can be tailored to target specific pollutants, making them a versatile choice for various industrial applications.

The objective of this chapter is to provide an overview of recent developments in the field of functional group polymers for wastewater treatment. We will discuss prospective uses of functional group polymers in water treatment, their benefits and downsides, the numerous functional group polymer types, their synthesis, and their effectiveness in eliminating various impurities from water. Recent developments in the use of functional group polymers for wastewater treatment are briefly covered in this chapter. Evaluation of the performance of functional group polymers, the impact they have on the environment, and upcoming advancements in this field will also be discussed. Additionally, it is important to gain a thorough understanding of the potential of functional group polymers for the design of cost-effective and efficient wastewater treatment techniques, as well as to make recommendations in this regard.

## 5.2 Functional Polymeric Materials Used in Wastewater Treatment

In the field of water treatment, functional polymeric materials have attracted a lot of attention. These substances may be natural or manufactured, and they may be altered to form different compounds (R. Hu et al., 2015). Numerous polymers are used in membrane fabrication as immobilization substrates and resins to target particular wastewater contaminants (Halake et al., 2014). Wood, shells, leaves, and seashells produce materials such as chitosan, cellulose, alginate, and cyclodextrins (Tran et al., 2015). In functional biomaterials, the interaction of adsorptive properties, functional groups, and pollutant accumulation is becoming more important. Furthermore, these polymeric materials have excellent durability and porosity, and their surface can be adjusted to enhance their interaction with pesticides. According to R. Zhao et al. (2017), the intrinsic process involves anionic contaminants adhering to the polymer's cationic amino group. The pKa of the molecule significantly impacts the adsorption of cationic pollutants such as pesticides because molecules with high pKa exhibit more dissociation, revealing negative charges that may interfere with the positive polymer (Rasheed et al., 2021).

## 5.2.1 Treatment Approaches

Ion exchange, coagulation/flocculation, adsorption, and membrane separation are used for the removal of heavy metals, organic pollutants, pharmaceuticals, and pesticides during wastewater treatment using various functional polymers and nanomaterials (NMs).

### 5.2.1.1 Heavy Metal Removal

Functional groups with anionic characteristics, such sulfonic acid and carboxylic acid groups, are widely employed to remove heavy metals from wastewater. Strong affinities exist between these functional groups and positively charged heavy metal ions. The functional groups that bind to heavy metal ions draw them out of the water. One example of a functional group employed for removing heavy metals is chitosan, a biopolymer with amino and hydroxyl functional groups. Chitosan can efficiently remove heavy metals from wastewater, including lead, copper, and chromium (Kyzas et al., 2014).

### 5.2.1.2 Organic Pollutant Removal

Furthermore, organic contaminants can be removed from wastewater using functional groups as a filter. One example is the usage of cationic functional groups, which can interact electrostatically to bind to organic contaminants. Ammonium and phosphonium groups are two other examples. Another example is the application of polymeric functional groups, such as chelating agents, which can form complexes with organic contaminants. Chelating agents are compounds containing two or more functional groups that may bind to a metal ion to form a ring structure. After that, the organic contaminants can attach to this ring structure and draw itself out of the water.

### 5.2.1.3 Elimination of Nutrients

Functional groups can also be used to remove nutrients, such as nitrogen and phosphorus, from wastewater. Phosphorus can be removed from anionic functional groups like sulfonic and carboxylic acids by ion exchange. Cationic functional groups like ammonium and phosphonium groups can be employed to remove nitrogen by biological or ion exchange processes. The functional groups, for example, can promote the development of bacteria that can convert nitrogen into nitrogen gas and subsequently release it into the atmosphere. From these, membrane separation and coagulation/flocculation approaches are the ones that are most commonly utilized in the industrial setting to treat industrial effluents.

## 5.2.2 Flocculation and Coagulation

Flocculation and coagulation are widely used to remove contaminants from wastewater. Coagulation causes the smaller particles to cluster and enlarge by neutralizing the charges on the suspended particles in wastewater with chemical agents. The larger particles can then be easily removed from the water. The process of flocculation, on the other hand, combines suspended particles using polymers to help agglomerate them and make the water purification process easier. For instance, a few polymers are utilized in the coagulation/flocculation technique, including polyacrylamide (PAM), poly(diallyldimethylammonium chloride) (PDADMAC), and epichlorohydrin/dimethylamine polymers.

One of the most widely used functional group polymers for coagulation and flocculation is PAM, a polymer that dissolves in water. PAM is particularly effective in flocculating and removing suspended particles from wastewater. PAM has been shown to be very efficient at removing impurities, such as suspended particles, organic compounds, and heavy metals, from industrial and municipal wastewater treatment systems (Peng et al., 2020).

Another functional group polymer widely used for coagulation and flocculation is PDADMAC. With the use of the cationic polymer PDADMAC, wastewater suspended particles can be more easily agglomerated and removed from the water by neutralizing the charges on those particles. The removal of a range of contaminants, such as suspended particles, organic compounds, and heavy metals, has been shown to be particularly effective when using PDADMAC (Zeng et al., 2014).

## 5.2.3 Various Functional Polymeric Materials for Membrane Separation

Specific conductive polymers such as polypyrrole, polyaniline, and polythiophene have some flexible properties including electrical conductivity and semiconductor-based characteristics that can be employed in membrane applications (Sarkar & Das, 2017). However, polymeric materials utilized in membranes have the problem of biofilm development, which leads to surface passivation. Materials with antibacterial properties, such as silver-loaded polymers, may prevent the formation of biofilms. Biomaterials are a low-cost alternative for generating flexible pesticide removal filters and membranes (A. Khan et al., 2018; Rasheed et al., 2021).

## 5.3 Types of Membrane Functional Polymers and Nanomaterials

### 5.3.1 Cationic Polymers and SiO$_2$ to Treat Pesticides in Wastewater

SiO$_2$ and cationic polymers are used for the treatment of pesticides in wastewater. Cationic polymers are long-chain polymers with positively charged groups along their length that can interact with negatively charged wastewater particles, including pesticides, in addition to SiO$_2$. The resulting composite material can efficiently adsorb and remove pesticides from wastewater when combined with SiO$_2$. Electrostatic attraction between the positively charged polymer and negatively charged pesticides is the mechanism of action of cationic polymers in the removal of pesticides from wastewater. The adsorption of pesticides onto the polymer surface is made possible by this electrostatic interaction, which effectively removes pesticides from wastewater.

$SiO_2$ is added to cationic polymers to increase their capacity for adsorption by creating a large surface area for interacting with pesticides. Additionally, $SiO_2$ nanoparticles (NPs) can act as nucleation sites for the polymer's growth, creating a more stable composite material. Adsorption of the pesticides onto the composite material, which effectively removes them from wastewater, is the general mechanism of action for cationic polymers and $SiO_2$ in the treatment of pesticides in wastewater. The high surface area of the $SiO_2$ component and electrostatic interactions between the polymer and pesticide drives this process (Dutta & Nath, 2018; Jin et al., 2012).

Tseng et al. (2015) conducted an experiment in which cationic polyacrylamide (CPAM) was used as the flocculant and silica microspheres as the coagulant. According to the findings, CPAM and silica microspheres together had a removal rate of up to 99% for pesticides from wastewater. Other than cationic polymers, sulfonated and anionic polymers are also used for the removal of pesticides from water bodies.

### 5.3.2 Sulfonated Polymers and SiO$_2$ to Treat Pesticides in Wastewater

Wastewater treated with sulfonated polymers and $SiO_2$ frequently contains pesticides. As flocculants, sulfonic polymers help the pesticide particles group together and form bigger clumps that can be more easily removed from the water. The addition of $SiO_2$, or silica, to sulfonic polymers can improve the flocculation process by increasing the surface area available for the pesticide particles to adhere to.

Various sulfonated polymers were used to remove pesticides from wastewater; the flocculant and coagulant approach uses sulfonated polyacrylamide (SPA) and $SiO_2$ NPs. According to the findings, SPA and silica NPs together had a removal rate of up to 98% for pesticides from wastewater. Generally speaking, sulfonated polymers and $SiO_2$ are efficient at removing pesticides from wastewater and have the potential to be used as a low-cost, environmentally friendly alternative to other methods of treating pesticides in wastewater (Căprărescu et al., 2021; Jin et al., 2012). Sulfonated polymers have also produced significant results in the pesticide removal process.

### 5.3.3 Hydroxyl Group-Functionalized Polymeric Membrane for Removal of Pesticides from Wastewater

Functionalization of a membrane using graphene oxide (GO) increases the relationship between pesticide removal and degradation time because the NM boosts catalytic activity by adding hydroxyl groups and hydrophilicity to the membrane, enhancing selectivity toward the pesticide (L. Zhao et al., 2020). Aside from GO, the popularity of carbon nanotubes (CNTs) has grown rapidly due to factors such as economy, mass production, a wide range of applications, and higher organic compound adsorption capacity, compared to other bulk materials (X. Hu et al., 2021; F. S. A. Khan et al., 2021). CNTs, unlike GOs, must be modified with strong acids to obtain carboxyl, hydroxyl, and amine groups, allowing for better incorporation into the membrane (Adamczak et al., 2021). Mechrez et al. (2014) fabricated an enzymatically active membrane functionalized with carboxylated carbon nanotubes (CNT-COOH) for the removal of methyl paraoxon; they attained 23% removal of the organophosphate compared to using the neat membrane, which can only remove 3%.

Adamczak et al. (2021) fabricated two functionalized PES membranes, a neat membrane and another with carboxylated CNTs (named PES@CNT-COOH), for the removal of endosulfan from water. When compared to the pristine membrane, the ratio of remotion for both functionalized membranes is 100%. PES@CNT-COOH completely eliminated the pollutant due to interactions generated by its hydroxyl groups, whereas PES@CNT removed endosulfan due to its high surface area of 1075 $m^2g^{-1}$ and smaller pore size.

Metal oxide NPs that have received the greatest attention are ZnO, $TiO_2$, $ZrO_2$, $SiO_2$, and $Fe_2O_3$ (Pordel et al., 2019; Vatanpour et al., 2022). In research studies including Zhao, Yan et al. (2015), ZnO was successfully embedded into polysulfone (PSF) and PES membranes. According to the studies of Shafiq et al. (2018) and Ren et al. (2017), $TiO_2$ was better integrated into PVDF membranes. Another approach for functionalizing cellulose acetate (CA) membranes is using $SiO_2$ (Ahmad et al., 2015).

Pordel et al. (2019) observed that the addition of $SiO_2$ and $Fe_2O_3$ NPs retained traces of atrazine and diazinon in a shorter filtration time (60 min), correlated to

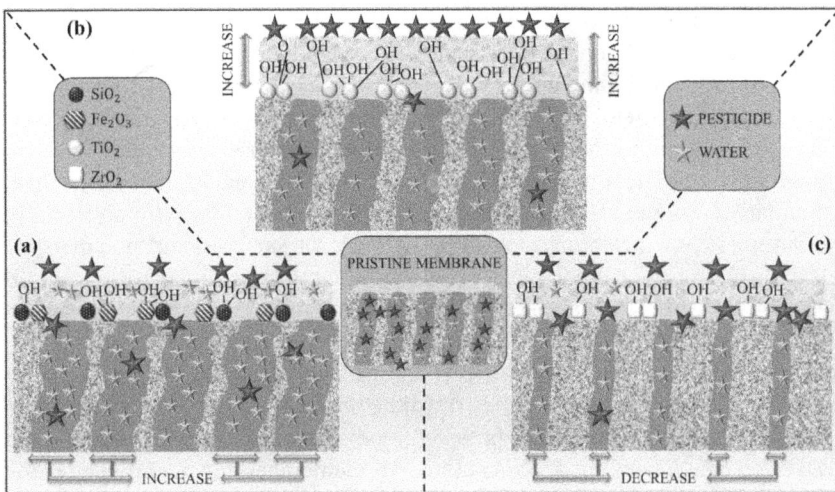

**FIGURE 5.2**
Comparison of functionalized membranes with $SiO_2$ and $Fe_2O_3$ NPs (a), $TiO_2$ NPs (b), and $ZrO_2$ NPs (c) (Sanoja-López et al., 2023). *Reproduced with permission from Elsevier.*

their neat PSF and polyacrylonitrile (PAN) membranes. $TiO_2$ NPs, on the other hand, considerably increase the antifouling capabilities of membranes because they contain a large concentration of hydroxyl groups, which allows for a greater affinity layer with water and hence reduces contaminant fouling on the membrane (Figure. 5.2). Song et al. (2016) showed that membranes functionalized with $TiO_2$ NPs have reduced fouling and pore size than pristine membranes and may be recycled 10-fold. However, among the metal oxides, $ZrO_2$ NPs are the most effective in decreasing pore size, which promotes an improvement in pesticide retention, with a reduction in permeability. Qin et al. (2020) achieved a $ZrO_2$ NP nanofiltration (NF) membrane with a reduced pore size. This enables the functionalized membrane to attain 89% removal of carbofuran.

Depending on the NMs employed to functionalize the membrane, $SiO_2$ and $Fe_2O_3$ provide enhanced pore size and permeability, $ZrO_2$ provides lower pore size, and $TiO_2$ provides greater antifouling properties.

### 5.3.4 Antibiotics Can Be Removed From Wastewater Using Cationic Functionalized Membranes

Palacio et al. (2020) performed a study with an ultrafiltration membrane conjugated with alkylated chitosan polyelectrolyte (ChA-PE) to remove antibiotics such as amoxicillin (AMX), ciprofloxacin (CIP), and tetracycline (TET) in aquatic ecosystems. Alkylated chitosan attained 94% efficiency, resulting in a dissolving polymer at the entire pH range. Maximum retention was achieved at pH 11 (70%), with electrostatic interactions between the antibiotic and polymer serving as the crucial mechanism. At a pH of 11, the ChA-PE demonstrated irreversibly bound antibiotic interaction values for CIP, AMX, and TET of 0.51, 0.74, and 0.92, respectively. This indicates that the polymer exhibits greater long-term interactions with AMX and TET. The ChA-PE, on the other hand, showed maximum retention capacities of 185.6, 420.2, and 632.8 mg $g^{-1}$ for CIP, AMX, and TET, respectively. This strongly indicates that hydrogen bonds exist in the system as a result of the donor–acceptor hydrogen bond groups that the antibiotics present. The association efficiency percentage values of CIP, AMX, and TET were 73.54, 87.08, and 93.83%, respectively.

Similarly, Wang et al. (2021) conducted an experiment with NF polyelectrolyte multilayer (PEM) membranes constructed with various cationic polymers, i.e., PDADMAC, and anionic polymers, i.e., poly(sodium styrenesulfonate) via layer-by-layer (LbL) assembly for the selective removal of antibiotics and salts from water bodies. The capacity of PEM and NF270 membranes to retain perfluoroalkyl compounds (such as perfluorooctanoic acid (PFOA) and perfluorooctanesulfonic acid (PFOS)) and antibiotics (such as amoxicillin trihydrate and tetracycline hydrochloride) was investigated using feed solutions prepared by dissolving 1 mg/L of each antibiotic in DI water and a combined salt solution. All membranes consistently had significant emerging organic contaminants (EOC) retentions (>85%), whereas the two-bilayer membrane had relatively ineffective removal efficiencies of between 60% and 70%. The EOCs examined in this study have comparably high molecular weights— between 414 and 538 g/mol—but they have various functional groups, which results in a diversity of charge characteristics. Due to their carboxylic and sulfonic acid head groups, respectively, PFOA and PFOS are both negatively charged at the pH of natural waters. The antibiotic molecules, in contrast, are amphoteric because they have functional groups with different values for the dissociation constant. In particular, amoxicillin trihydrate has carboxyl (pKa1 = 2.7), amino (pKa2 = 7.5), and hydroxyl (pKa3 = 9.6) groups (Putra et al., 2009), whereas tetracycline hydrochloride has tricarbonyl (pKa1 = 3.2–3.3), phenolic diketone (pKa2 = 7.3–7.7), and dimethylamine (pKa3 = 9.1–9.7) groups (Figure 5.3) (Pollard & Morra, 2018). The four- and six-layer membranes showed reasonably good retention of all EOCs (>85%, green and blue bars in Figure 5.4), comparable to the NF270 membrane, while having distinct charge characteristics. This result indicates that regardless of their charge characteristics, four- and six-bilayer PEMs reject EOCs primarily through size exclusion. Contrarily, electrostatic repulsive interactions between the negatively charged EOCs (such as PFOA and PFOS) and the PEM surface hindered EOC transport across the membrane, and the very thin and sparse two-bilayer membrane did not serve as an effective barrier to the passage of EOCs.

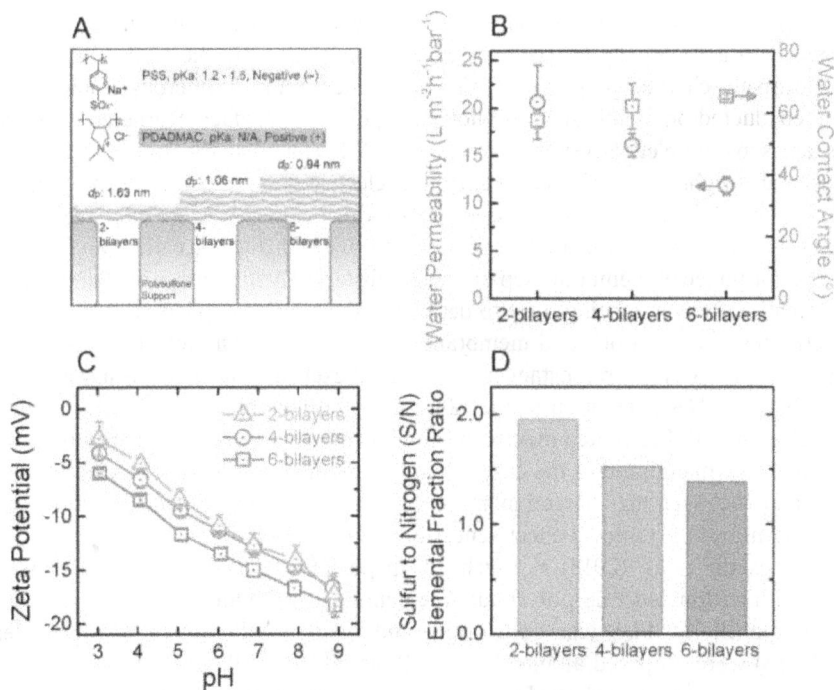

**FIGURE 5.3**

(a) Schematic representation of polyelectrolyte multilayer (PEM) NF membranes fabricated via LbL assembly of a cationic polymer, i.e., PDADMAC, and an anionic polymer, i.e., PSS. (b) Water permeability, (c) zeta potential of the membranes, and (d) sulfur/nitrogen ratio of PEM NF LbL membranes (Wang et al., 2021). *Reproduced with permission from the American Chemical Society.*

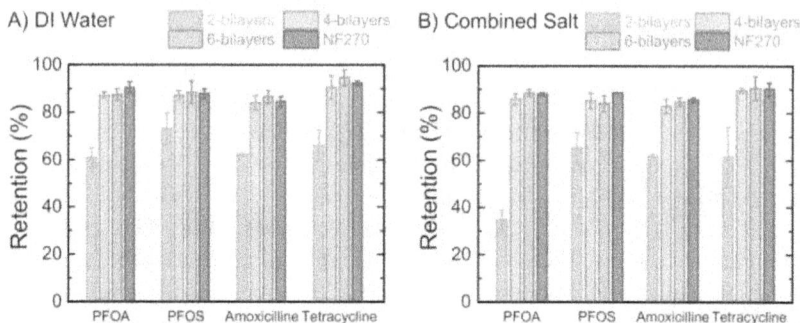

**FIGURE 5.4**
Retention percentage of perfluoroalkyl and antibiotic pollutants on various LbL PEMs (Wang et al., 2021). *Reproduced with permission from the American Chemical Society.*

### 5.3.5 Sulfonated Polymers for the Treatment of Antibiotics in Wastewater

Zhao et al. (2015) conducted a study with a composite NF membrane with the surface coated with a negatively charged sulfonated poly(ether ether ketone) (SPEEK) on the positively charged polyethylenimine (PEI) membrane. Numerous studies have been conducted on membrane morphologies, pore size, surface charge characteristics, and separation effectiveness. While the SPEEK-coated membrane had a reduced molecular weight cut-off (MWCO) and an isoelectric point of 5.4, the PEI-modified membrane exhibited a positive charge at pH levels below 9.8. More effectively than monovalent ions, SPEEK-coated membranes remove both divalent cations and anions. The performance of membrane separation is substantially influenced by solution pH. Systematic research is being done to determine the relationship between separation effectiveness, solution pH, and membrane surface charge. The results demonstrate that under various pH circumstances, the Donnan exclusion mechanism significantly contributes to NaCl retention of the SPEEK-coated membrane. Due to its smaller pore size, the SPEEK-coated membrane exhibits greater removal efficiency than the PEI-modified membrane in the case of sulfamerazine separation. The electrostatic interaction between the charged membrane and the sulfamerazine dissociation species determines the removal effectiveness in various pH solutions.

TheRostam et al. (2018) research group performed an antibacterial experiment with sulfonated PES/polyrhodanine nanocomposite membrane. To increase its hydrophilicity, PES was sulfonated, and then polyrhodanine nanoparticles (PRhNPs) were prepared alongside the sulfonated PES (SPES) by polyrhodanine (PRh) *in situ* polymerization. Due to the development of potential bondings and polymer engagements, the sulfonation process also aids in the making of composite membranes. Improved hydrophilicity in the SPES/PRh membrane led to better fluxes for aqueous solutions. The composite SPES/PRh membrane flow was increased from 58.21 L/m² h for SPES to 139/78 L/m² h. The SPES/PRh membrane demonstrated enhanced flow, better rejection, and adequate antibacterial and antibiofouling capabilities in membrane operating performances, antibacterial activity, and antibiofouling testing. For the recommended concentrations and appropriate inhibitory zones up to 9 mm, 100% bacteria mortality has been accomplished.

### 5.3.6 Hydroxyl Group-Incorporated Polymeric Membranes to Treat Antibiotics in Wastewater

Shakak et al. (2020) performed a study with PSF/polyvinylpyrrolidone/SiO$_2$ nanocomposite ultrafiltration membranes for the removal of AMX from water bodies. In the range of 0–4 wt%, SiO$_2$ NPs (10–20 nm) were employed as a hydrophilic agent to modify the ultrafiltration membrane surface. Additionally, the efficacy of AMX separation, pure water flux, and an analysis of the fouling characteristics were used to estimate the effectiveness of the synthesized membranes. The hydroxyl group in the SiO$_2$ structure greatly boosts the hydrophilicity of nanocomposite membranes. With more SiO$_2$ in the polymer solution, the surface roughness is significantly increased. The modified membranes showed noticeable improvements in antifouling properties. The membrane flux increased as the structure's SiO$_2$ concentration changed: 6.6 L/m²/h for the unmodified membrane to 42.28 L/m²/h for the membrane containing 3 wt% of NPs. Investigation of membrane performance revealed that the AMX separation performance increased from 66.52% to 89.81% with an increase in SiO$_2$ NPs from 0 to 4 wt%. The relative flux reduction ratio (RFR), which was 57.03% for the unmodified membrane but reduced to 22.89% for the modified membrane with 3 wt% of NPs, also improved by the addition of SiO$_2$ NPs, according to the analysis of the membrane fouling parameters.

## 5.4 Potential and Challenges

In conclusion, because of the distinct physiochemical characteristics of polymers and NMs, functional groups have been introduced to enhance their performance in wastewater treatment. The removal of organic pollutants and nutrients from wastewater has been accomplished using cationic, sulfonated, hydroxyl, and amine group-based functionalized polymeric materials. They are also used in applications for the removal of pesticides and antibiotics from the aquatic ecosystem.

The ability of functional groups to remove a variety of contaminants from wastewater is one of the main advantages of using them in wastewater treatment. For instance, sulfonic acid and hydroxyl and amino groups are efficient at removing organic pollutants from wastewater through functionalized membranes.

In conclusion, functional groups have already demonstrated a lot of potential for wastewater treatment, and it is likely that they will continue to play a big part in upcoming technologies. By utilizing the unique characteristics of functional groups, we can improve the efficacy and efficiency of wastewater treatment and contribute to a more sustainable and clean future.

## References

Adamczak, M., Kamińska, G., & Bohdziewicz, J. (2021). Relationship between the addition of carbon nanotubes and cut-off of ultrafiltration membranes and their effect on retention of microcontaminants. *Desalination and Water Treatment, 214*, 263–272. https://doi.org/10.5004/dwt.2021.26698

Ahmad, A., Waheed, S., Khan, S. M., e-Gul, S., Shafiq, M., Farooq, M., Sanaullah, K., & Jamil, T. (2015). Effect of silica on the properties of cellulose acetate/polyethylene glycol membranes for reverse osmosis. *Desalination, 355*, 1–10. https://doi.org/10.1016/j.desal.2014.10.004

Al-Hamdi, A. M., Sillanpää, M., Bora, T., & Dutta, J. (2016). Efficient photocatalytic degradation of phenol in aqueous solution by $SnO_2$:Sb nanoparticles. *Applied Surface Science, 370*, 229–236. https://doi.org/10.1016/j.apsusc.2016.02.123

Arumugham, T., Kaleekkal, N. J., Gopal, S., Nambikkattu, J., K, R., Aboulella, A. M., Ranil Wickramasinghe, S., & Banat, F. (2021). Recent developments in porous ceramic membranes for wastewater treatment and desalination: A review. *Journal of Environmental Management, 293*, 112925. https://doi.org/10.1016/j.jenvman.2021.112925

Bapat, G., Mulla, J., Labade, C., Ghuge, O., Tamhane, V., & Zinjarde, S. (2022). Assessment of recombinant glutathione-S-transferase (HaGST-8) silica nano-conjugates for effective removal of pesticides. *Environmental Research, 204*, 112052. https://doi.org/10.1016/j.envres.2021.112052

Bilici, Z., Ozay, Y., Yuzer, A., Ince, M., Ocakoglu, K., & Dizge, N. (2021). Fabrication and characterization of polyethersulfone membranes functionalized with zinc phthalocyanines embedding different substitute groups. *Colloids and Surfaces A: Physicochemical and Engineering Aspects, 617*, 126288. https://doi.org/10.1016/j.colsurfa.2021.126288

Căprărescu, S., Modrogan, C., Purcar, V., Dăncilă, A. M., & Orbuleț, O. D. (2021). Study of polyvinyl alcohol-$SiO_2$ nanoparticles polymeric membrane in wastewater treatment containing zinc ions. *Polymers, 13*(11), Article 11. https://doi.org/10.3390/polym13111875

Chen, C., Luo, J., Bu, C., Zhang, W., & Ma, L. (2022). Efficacy of a large-scale integrated constructed wetland for pesticide removal in tail water from a sewage treatment plant. *Science of the Total Environment, 838*, 156568. https://doi.org/10.1016/j.scitotenv.2022.156568

Dutta, D. P., & Nath, S. (2018). Low cost synthesis of $SiO_2$/C nanocomposite from corn cobs and its adsorption of uranium (VI), chromium (VI) and cationic dyes from wastewater. *Journal of Molecular Liquids, 269*, 140–151. https://doi.org/10.1016/j.molliq.2018.08.028

Halake, K., Birajdar, M., Kim, B. S., Bae, H., Lee, C., Kim, Y. J., Kim, S., Kim, H. J., Ahn, S., An, S. Y., & Lee, J. (2014). Recent application developments of water-soluble synthetic polymers. *Journal of Industrial and Engineering Chemistry, 20*(6), 3913–3918. https://doi.org/10.1016/j.jiec.2014.01.006

Hassaan, M. A., & El Nemr, A. (2020). Pesticides pollution: Classifications, human health impact, extraction and treatment techniques. *The Egyptian Journal of Aquatic Research, 46*(3), 207–220. https://doi.org/10.1016/j.ejar.2020.08.007

Hu, R., Dai, S., Shao, D., Alsaedi, A., Ahmad, B., & Wang, X. (2015). Efficient removal of phenol and aniline from aqueous solutions using graphene oxide/polypyrrole composites. *Journal of Molecular Liquids, 203*, 80–89. https://doi.org/10.1016/j.molliq.2014.12.046

Hu, X., You, S., Li, F., & Liu, Y. (2021). Recent advances in antimony removal using carbon-based nanomaterials: A review. *Frontiers of Environmental Science & Engineering, 16*(4), 48. https://doi.org/10.1007/s11783-021-1482-7

Jin, L. M., Yu, S. L., Shi, W. X., Yi, X. S., Sun, N., Ge, Y. L., & Ma, C. (2012). Synthesis of a novel composite nanofiltration membrane incorporated $SiO_2$ nanoparticles for oily wastewater desalination. *Polymer*, *53*(23), 5295–5303. https://doi.org/10.1016/j. polymer.2012.09.014

Khan, A., Khan, A. A. P., Rahman, M. M., Asiri, A. M., Inamuddin, Alamry, K. A., & Hameed, S. A. (2018). Preparation and characterization of PANI@G/CWO nanocomposite for enhanced 2-nitrophenol sensing. *Applied Surface Science*, *433*, 696–704. https://doi. org/10.1016/j.apsusc.2017.09.219

Khan, F. S. A., Mubarak, N. M., Khalid, M., Tan, Y. H., Abdullah, E. C., Rahman, M. E., & Karri, R. R. (2021). A comprehensive review on micropollutants removal using carbon nanotubes-based adsorbents and membranes. *Journal of Environmental Chemical Engineering*, *9*(6), 106647. https://doi.org/10.1016/j.jece.2021.106647

Kyzas, G. Z., Siafaka, P. I., Lambropoulou, D. A., Lazaridis, N. K., & Bikiaris, D. N. (2014). Poly (itaconic acid)-grafted chitosan adsorbents with different cross-linking for Pb (II) and Cd (II) uptake. *Langmuir*, *30*(1), 120–131. https://doi.org/10.1021/ la402778x

Lee, J. K. Y., Chen, N., Peng, S., Li, L., Tian, L., Thakor, N., & Ramakrishna, S. (2018). Polymer-based composites by electrospinning: Preparation & functionalization with nanocarbons. *Progress in Polymer Science*, *86*, 40–84. https://doi.org/10.1016/j. progpolymsci.2018.07.002

Lin, Y.-S., Wu, S.-H., Hung, Y., Chou, Y.-H., Chang, C., Lin, M.-L., Tsai, C.-P., & Mou, C.-Y. (2006). Multifunctional composite nanoparticles: Magnetic, luminescent, and mesoporous. *Chemistry of Materials*, *18*(22), 5170–5172. https://doi.org/10.1021/ cm061976z

Mechrez, G., Krepker, M. A., Harel, Y., Lellouche, J.-P., & Segal, E. (2014). Biocatalytic carbon nanotube paper: A 'one-pot' route for fabrication of enzyme-immobilized membranes for organophosphate bioremediation. *Journal of Materials Chemistry B*, *2*(7), 915–922. https://doi.org/10.1039/C3TB21439G

Mohmood, I., Lopes, C. B., Lopes, I., Ahmad, I., Duarte, A. C., & Pereira, E. (2013). Nanoscale materials and their use in water contaminants removal—A review. *Environmental Science and Pollution Research*, *20*(3), 1239–1260. https://doi.org/10.1007/s11356-012-1415-x

Palacio, D. A., Becerra, Y., Urbano, B. F., & Rivas, B. L. (2020). Antibiotics removal using a chitosan-based polyelectrolyte in conjunction with ultrafiltration membranes. *Chemosphere*, *258*, 127416. https://doi.org/10.1016/j.chemosphere.2020.127416

Pejman, M., Dadashi Firouzjaei, M., Aghapour Aktij, S., Zolghadr, E., Das, P., Elliott, M., Sadrzadeh, M., Sangermano, M., Rahimpour, A., & Tiraferri, A. (2021). Effective strategy for UV-mediated grafting of biocidal Ag-MOFs on polymeric membranes aimed at enhanced water ultrafiltration. *Chemical Engineering Journal*, *426*, 130704. https://doi. org/10.1016/j.cej.2021.130704

Peng, Y., Jin, D., Li, J., & Wang, C. (2020). Flocculation of mineral processing wastewater with Polyacrylamide. *IOP Conference Series: Earth and Environmental Science*, *565*(1), 012101. https://doi.org/10.1088/1755-1315/565/1/012101

Pollard, A. T., & Morra, M. J. (2018). Fate of tetracycline antibiotics in dairy manure-amended soils. *Environmental Reviews*, *26*(1), 102–112.

Pordel, M. A., Maleki, A., Ghanbari, R., Rezaee, R., Khamforoush, M., Daraei, H., Athar, S. D., Shahmoradi, B., Safari, M., Ziaee, A. H., Lalhmunsiama, & Lee, S.-M. (2019). Evaluation of the effect of electrospun nanofibrous membrane on removal of diazinon from aqueous solutions. *Reactive and Functional Polymers*, *139*, 85–91. https://doi. org/10.1016/j.reactfunctpolym.2019.03.017

Putra, E. K., Pranowo, R., Sunarso, J., Indraswati, N., & Ismadji, S. (2009). Performance of activated carbon and bentonite for adsorption of amoxicillin from wastewater: Mechanisms, isotherms and kinetics. *Water Research, 43*(9), 2419–2430. https://doi.org/10.1016/j.watres.2009.02.039

Qin, H., Guo, W., Huang, X., Gao, P., & Xiao, H. (2020). Preparation of yttria-stabilized ZrO2 nanofiltration membrane by reverse micelles-mediated sol-gel process and its application in pesticide wastewater treatment. *Journal of the European Ceramic Society, 40*(1), 145–154. https://doi.org/10.1016/j.jeurceramsoc.2019.09.023

Rana, A. K., Mishra, Y. K., Gupta, V. K., & Thakur, V. K. (2021). Sustainable materials in the removal of pesticides from contaminated water: Perspective on macro to nanoscale cellulose. *Science of the Total Environment, 797*, 149129. https://doi.org/10.1016/j.scitotenv.2021.149129

Rasheed, T., Rizwan, K., Bilal, M., Sher, F., & Iqbal, H. M. N. (2021). Tailored functional materials as robust candidates to mitigate pesticides in aqueous matrices—A review. *Chemosphere, 282*, 131056. https://doi.org/10.1016/j.chemosphere.2021.131056

Ren, L., Tao, J., Chen, H., Bian, Y., Yang, X., Chen, G., Zhang, X., Liang, G., Wu, W., Song, Z., & Wang, Y. (2017). Myeloid differentiation protein 2-dependent mechanisms in retinal ischemia-reperfusion injury. *Toxicology and Applied Pharmacology, 317*, 1–11. https://doi.org/10.1016/j.taap.2017.01.001

Rostam, A. B., Peyravi, M., Ghorbani, M., & Jahanshahi, M. (2018). Antibacterial surface modified of novel nanocomposite sulfonated polyethersulfone/polyrhodanine membrane. *Applied Surface Science, 427*, 17–28. https://doi.org/10.1016/j.apsusc.2017.08.025

Sanoja-López, K. A., Quiroz-Suárez, K. A., Dueñas-Rivadeneira, A. A., Maddela, N. R., Montenegro, M. C. B. S. M., Luque, R., & Rodríguez-Díaz, J. M. (2023). Polymeric membranes functionalized with nanomaterials (MP@NMs): A review of advances in pesticide removal. *Environmental Research, 217*, 114776. https://doi.org/10.1016/j.envres.2022.114776

Sarkar, S., & Das, R. (2017). PVP capped silver nanocubes assisted removal of glyphosate from water—A photoluminescence study. *Journal of Hazardous Materials, 339*, 54–62. https://doi.org/10.1016/j.jhazmat.2017.06.014

Shafiq, M., Sabir, A., Islam, A., Khan, S. M., Gull, N., Hussain, S. N., & Butt, M. T. Z. (2018). Cellulaose acetate based thin film nanocomposite reverse osmosis membrane incorporated with TiO2 nanoparticles for improved performance. *Carbohydrate Polymers, 186*, 367–376. https://doi.org/10.1016/j.carbpol.2018.01.070

Shakak, M., Rezaee, R., Maleki, A., Jafari, A., Safari, M., Shahmoradi, B., Daraei, H., & Lee, S.-M. (2020). Synthesis and characterization of nanocomposite ultrafiltration membrane (PSF/PVP/SiO2) and performance evaluation for the removal of amoxicillin from aqueous solutions. *Environmental Technology & Innovation, 17*, 100529. https://doi.org/10.1016/j.eti.2019.100529

Song, Z., Fathizadeh, M., Huang, Y., Chu, K. H., Yoon, Y., Wang, L., Xu, W. L., & Yu, M. (2016). TiO$_2$ nanofiltration membranes prepared by molecular layer deposition for water purification. *Journal of Membrane Science, 510*, 72–78. https://doi.org/10.1016/j.memsci.2016.03.011

Tran, V. S., Ngo, H. H., Guo, W., Zhang, J., Liang, S., Ton-That, C., & Zhang, X. (2015). Typical low cost biosorbents for adsorptive removal of specific organic pollutants from water. *Bioresource Technology, 182*, 353–363. https://doi.org/10.1016/j.biortech.2015.02.003

Tseng, S., Hsu, Y.-R., & Hsu, J.-P. (2015). Diffusiophoresis of polyelectrolytes: Effects of temperature, pH, type of ionic species and bulk concentration. *Journal of Colloid and Interface Science, 459*, 167–174. https://doi.org/10.1016/j.jcis.2015.08.014

Vatanpour, V., Jouyandeh, M., Akhi, H., Mousavi Khadem, S. S., Ganjali, M. R., Moradi, H., Mirsadeghi, S., Badiei, A., Esmaeili, A., Rabiee, N., Habibzadeh, S., Koyuncu, I., Nouranian, S., Formela, K., & Saeb, M. R. (2022). Hyperbranched polyethylenimine functionalized silica/polysulfone nanocomposite membranes for water purification. *Chemosphere, 290,* 133363. https://doi.org/10.1016/j.chemosphere.2021.133363

Wang, Y., Zucker, I., Boo, C., & Elimelech, M. (2021). Removal of emerging wastewater organic contaminants by polyelectrolyte multilayer nanofiltration membranes with tailored selectivity. *ACS ES&T Engineering, 1*(3), 404–414. https://doi.org/10.1021/acsestengg.0c00160

Zeng, T., Pignatello, J. J., Li, R. J., & Mitch, W. A. (2014). Synthesis and application of a quaternary phosphonium polymer coagulant to avoid n-nitrosamine formation. *Environmental Science & Technology, 48*(22), 13392–13401. https://doi.org/10.1021/es504091s

Zhao, L., Deng, C., Xue, S., Liu, H., Hao, L., & Zhu, M. (2020). Multifunctional g-C3N4/Ag NPs intercalated GO composite membrane for SERS detection and photocatalytic degradation of paraoxon-ethyl. *Chemical Engineering Journal, 402,* 126223. https://doi.org/10.1016/j.cej.2020.126223

Zhao, R., Li, X., Sun, B., Ji, H., & Wang, C. (2017). Diethylenetriamine-assisted synthesis of amino-rich hydrothermal carbon-coated electrospun polyacrylonitrile fiber adsorbents for the removal of Cr(VI) and 2,4-dichlorophenoxyacetic acid. *Journal of Colloid and Interface Science, 487,* 297–309. https://doi.org/10.1016/j.jcis.2016.10.057

Zhao, S., Yan, W., Shi, M., Wang, Z., Wang, J., & Wang, S. (2015). Improving permeability and antifouling performance of polyethersulfone ultrafiltration membrane by incorporation of ZnO-DMF dispersion containing nano-ZnO and polyvinylpyrrolidone. *Journal of Membrane Science, 478,* 105–116. https://doi.org/10.1016/j.memsci.2014.12.050

Zhao, S., Yao, Y., Ba, C., Zheng, W., Economy, J., & Wang, P. (2015). Enhancing the performance of polyethylenimine modified nanofiltration membrane by coating a layer of sulfonated poly(ether ketone) for removing sulfamerazine. *Journal of Membrane Science, 492,* 620–629. https://doi.org/10.1016/j.memsci.2015.03.017

# 6

# Functional Nanomaterials and Polymers for Desalination

## 6.1 Brief Overview of Desalination and Its Processes

Only 3% of water on the Earth is freshwater, and a little fraction of that is easily accessible soft water. Water covers 70% of the Earth's surface, much of it being unsuitable for domestic use. This is due to the fact that the majority of fresh water is gathered in either subsurface reservoirs or frozen glaciers (Teow & Mohammad, 2019).

To overcome this, conventional water treatment technologies can be used, such as primary (filtration, screening, separation, centrifugation, coagulation and flocculation, and sedimentation), secondary (anaerobic and aerobic treatments), and tertiary (crystallization, distillation, solvent extraction, evaporation, precipitation, oxidation, ion exchange, microfiltration (MF), ultrafiltration (UF), reverse osmosis (RO), nanofiltration (NF), adsorption, electrolysis, and electrodialysis) level water treatment technologies.

Due to the rising demand from population expansion and the decreasing desalination costs, desalination is becoming more competitive than traditional water treatment for urban use. The two primary forms of desalination technology are thermal-based and membrane-based technologies (Figure 6.1).

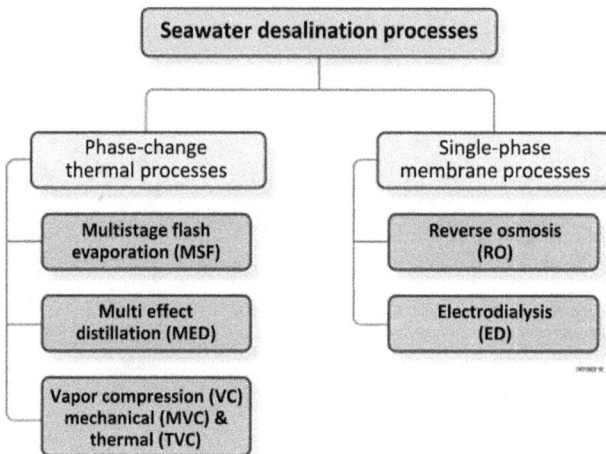

**FIGURE 6.1**
Major desalination processes (Al-Karaghouli et al., 2009). *Reproduced with permission from Elsevier.*

 DOI: 10.12019781003391364-6

Membrane surface modification and functionalization are crucial for controlling and modifying surface properties as well as for giving the membranes new functionalities including hydrophilicity and surface charge (Siddeeg et al., 2021).

Some techniques can effectively remove nutrients and organic debris but are inefficient at reducing salinity. They are also inappropriate for usage on a commercial scale (Das et al., 2014). Desalination is known to be one of the most promising sustainable methods for water treatment to supply fresh water (Y. Wang et al., 2017).

In this platform, membrane materials play an important role. Carbon structures with different sizes such as metal–organic frameworks (MOFs), carbon nanotubes (CNTs), covalent triazine frameworks (CTFs), and graphene were studied as promising nanomaterials (Y. Wang et al., 2017). Also, it has an effective strategy to remove a wide range of contaminants from water. For instance, functionalization of nanoporous graphene sheets with various active groups and inorganic nanoparticles enhances the desalination capabilities of pure graphene (Y. Wang et al., 2017).

Materials smaller than 100 nm in at least one dimension are often referred to as nanomaterials. A selective layer on the surface of a microporous substrate makes up the composite or asymmetric NF membrane. According to the surface charge properties, nanofiltration membranes (NFMs) can generally be separated into neutral membranes, positively charged membranes, and negatively charged membranes. The majority of commercial interfacially polymerized NFMs have neutral or negative charge characteristics (X. Wang et al., 2019).

Seawater desalination is carried out using nanomaterials including CNTs, aquaporin (AQP), carbon-based nanoparticles, inorganic nanoparticles, MOFs, graphene, and covalent organic frameworks (COFs) (Figure 6.2). These nanomaterials are frequently utilized in the membrane for membrane distillation (MD), reverse osmosis (RO), NF, forward osmosis (FO), and capacitive deionization (CDI) (Wei et al., 2021). The fouling and pore clogging caused by pollutant precipitation decrease the membrane's lifetime and module (Das et al., 2014).

## 6.2　Types of Functionalized Nanomaterials

### 6.2.1　Functionalized Silica

The existence of some silica nanoparticles (SNs) with diameters 50 and 100 nm on the thin-film layer of SN-thin-film composite (TFC) membranes shows that functionalized SNs were successfully integrated into the polyamide (PA) layers. Cross-sectional SEM images of the TFC and SN-TFC membranes, which have a maximum PA layer thickness of 450 nm, show the structure of the various membranes. The equilibrium contact angle decreases for SN-TFC membranes with amine and epoxy functionalization. Similar contact angle profiles are reported for all SN-TFC membranes.

Interestingly, the high hydrophilicity of TFC and thin-film nanocomposite (TFN) membranes is associated with enhanced water flux during the desalination process. As a result, the decreased contact angles of all SN-TFC membranes compared to

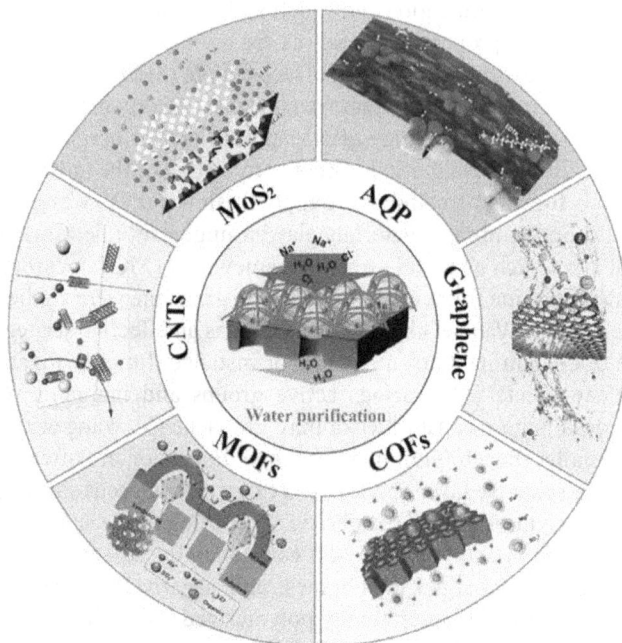

**FIGURE 6.2**
Functionalized nanomaterials used for MD (Wei et al., 2021). *Reproduced with permission from Elsevier.*

TFC membranes appear to enhance their desalination properties. Similar high water permeability values are seen as the integrated material's SN content increases from 0.05 to 0.1 wt%. The bulk PA layer's structure may change as a result of the addition of SNs, and the amount of free volume might increase as well.

However, after the addition of SN, the hydrophilicity of the TFC membranes rises, further increasing the solubilization and diffusion of water. Although it is evident that salt rejection values improve with concentration, SN-TFC membranes fare slightly worse than TFC membranes in this regard. The performance of the 0.05 wt% SN50-TFC membrane shows a negligibly low (only 0.7% lower, p = 0.7472) drop in water flux and a negligibly low (p = 0.9932) change in salt rejection when compared to the 0.1 wt% SN50-TFC membrane (Zargar et al., 2017; Baig et al., 2019).

To improve the hydrogen bonding sites and subsequently lessen the potential to foul while enhancing hydrophilicity, Tiraferri et al. (2012) functionalized the surface of SNs with super hydrophilic ligands. The average surface roughness (Ra) of the TFN membranes increased as the concentration of nano-Si increased, regardless of the solution they were dispersed in. The average surface roughness of both types of TFN membranes was 2.0 wt% nano-Si, which was higher than 0.05 wt%.

Pang and Zhang (2018) fabricated hydrophobic fluorinated silica/PA TFN films, and the results showed that the hydrophobic functionalization of SNs increased their dispersion in the organic solvent during the interfacial polymerization (IP) process. As a result, the ability of PA layers to integrate nanoparticles was improved, which improved the membrane's ability to reject salt while maintaining its water permeability (Abdelsamad et al., 2018).

MSN is believed to have an isoelectric point (IEP) at pH 3.7 ± 0.2. According to reports, the IEP of polyamide produced from m-phenylenediamine (MPD) and TMC is similar to that of MSN. The results demonstrate that there is no significant change in the IEP and only a slight reduction in the absolute zeta potential of the TFC membrane with an increase in the amount of octadecyltrichlorosilane (OTS)-functionalized silica.

The contact angle for TFN membranes increased from 68.4° to 84.4° when nanoparticle loading was raised from 0.003 to 0.05 w/v%, showing that the surface of TFN membranes becomes more hydrophobic. The TFC membrane has a contact angle of 67.3°. The loading of nanoparticles during the IP process determines how well they are incorporated into the PA thin layer (Figure 6.3).

## 6.2.2 Functionalized Aluminum

TFC, CA-TFN, and CO-TFN NF membranes (commercial-$Al_2O_3$ (CO-TFN) and camphor-$Al_2O_3$ NPs (CA-TFN)) are used in desalination. The CA-TFN (160 nm),

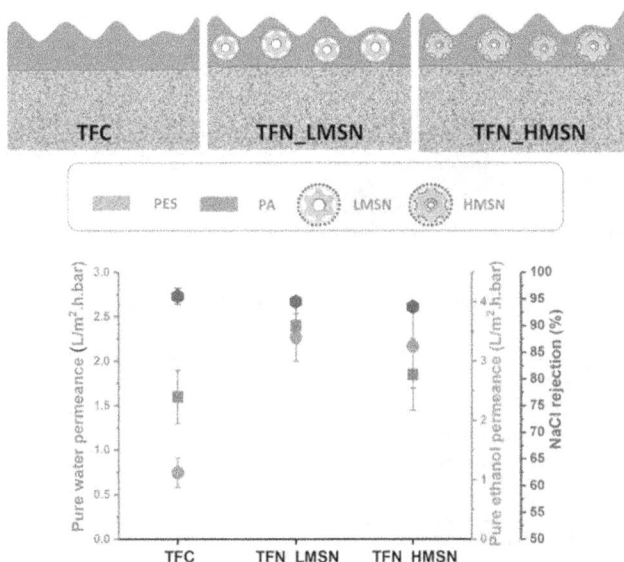

**FIGURE 6.3**
TFC of hydrophilic mesoporous silica nanoparticles (LMSNs) and hydrophobic mesoporous silica nanoparticles (HMSNs) for desalination (Abdelsamad et al., 2018). *Reproduced with permission from Elsevier.*

CO-TFN (190 nm), and TFC (240 nm) in that order can be used to rearrange the PA layer thickness in TFC, CO-TFN, and CA-TFN NF membranes. The salt retention percentages of $Na_2SO_4$ and NaCl were determined to be 96.5% and 95.1%, respectively, when 0.98 mM of commercial- and camphor-$Al_2O_3$ NPs were incorporated into the PA matrix. Additionally, increased from 0.98 mM to 2.94 mM when loading various NPs (Camphor- $Al_2O_3$ and commercial-$Al_2O_3$). For CA- $Al_2O_3$ membranes and CO-$Al_2O_3$ membranes, the rejection rate for correspondence decreased to 88.49% and 83.6%, respectively.

The flux (L/m² hr) increased in the following order: CA.TFN> CO.TFN> TFC. Enhancing the $Al_2O_3$/PA interfaces and including $Al_2O_3$ NPs increased flow mostly by increasing membrane free-volume and creating additional passageways for $H_2O$ molecules to enter membrane conformation through their internal nanopores. Furthermore, the salt rejection in TFC, CA.TFN, and CO.TFN membranes in the following order: $Na_2SO_4$> $MgSO_4$> NaCl> $MgCl_2$. The molecular weight of $Na_2SO_4$, $MgSO_4$, $Mg(HCO_3)_2$, and $Ca(HCO_3)_2$ is greater than or almost equal to 120, whereas the molecular weight of sodium chloride is 58.50 (Kotp, 2021). According to Huang et al. (2007), the polyols from the dried *C. camphora* leaf extract contributed to the decrease in aluminum ions when the band at 1024 cm$^{-1}$ in camphor NP-$Al_2O_3$ disappeared (Figure 6.4). Furthermore, Yang et al. (2010) suggested that the polyols present in broth as flavones, terpenoids, and polysaccharides are crucial in decreasing and stabilizing NPs (Kotp, 2021).

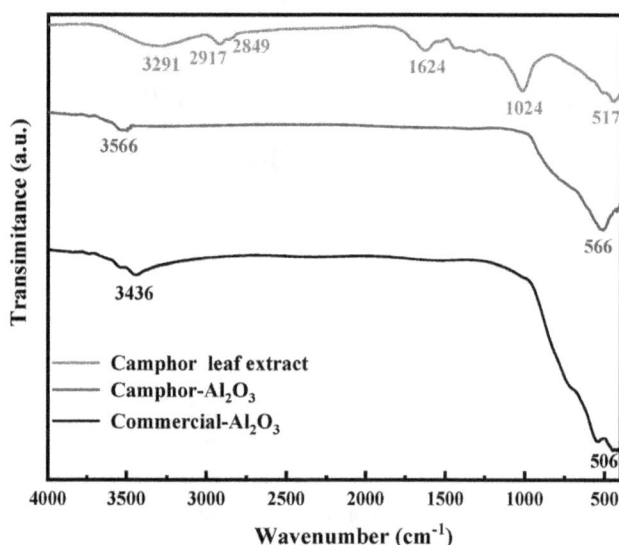

**FIGURE 6.4**
Functional group studies of unmodified $Al_2O_3$ and modified $Al_2O_3$ using camphor for MD (Kotp, 2021).
*Reproduced with permission from Elsevier.*

### 6.2.3 Functionalized TiO$_2$

As a result of numerous nanorods being embedded in polymeric membranes, a research group successfully synthesized the hierarchical structure of TiO$_2$ micro-flowers, which significantly improved their hydrophobic qualities. It has been suggested that combining micro- and nanostructures can boost the overall hydro-phobicity of the membrane surface. The membrane's strong resistance to liquid penetration is primarily caused by the ability of the combination to gather more air.

Using 1000 mg/L humic acid solution and ethylene glycol, the 1H,1H,2H,2H-perfluorodecyltriethoxysilane (C8)-flower-like/TiO$_2$-hollow fiber membrane (HFM) exhibited a superomniphobic surface (>150°), with corresponding contact angles of 160.9°±1.5 and 150°. This occurred after C8 and TiO$_2$ micro-flower function-alization. Olive oil has a contact angle with the membrane of 140.1°±1.6, making it almost superomniphobic. In comparison, the C8-HFM and C8-rod-like/TiO$_2$-HFM have decreased omniphobicity and have contact angles with olive oil that are, respectively, 104.7°±3.4 and 120°±2.5 (Abd Aziz et al., 2020).

### 6.2.4 Fluorine-Functionalized

For nanoporous graphene oxide (NPGO) and nanoporous graphene (NPG) films (fluorinated nanoporous graphene oxide (F-NPGO) and nanoporous graphene (F-NPG)), under operational conditions, a portion of the system's input water is taken as a concentrate or retentate water based on the recovery rate, and the salt rejection rate is then examined. Commercial RO systems have a recovery rate of between 30 and 50%. Recovery rate is calculated using the permeate water to feed water ratio. At all pressures, the F-NPGO membrane exhibits a 100% rejection rate. The salt rejection decreases with increasing applied pressure and pore size until it reaches a minimum value of 90.9% for F-NPG pores at 150 MPa. Ions can pass through large pores (F-NPGO and F-NPG), lessening salt rejection as pressure increases (Hosseini et al., 2019).

Since the fluorinated (-F) hole has more hydrophilic characteristics than the hydroxylated (-OH) pore, it is obvious that it can increase water permeability more. In this work, the water permeability and salt rejection percentage for the G2 system (with -F chemical function and 14.6 ± 2 pore size) are 3630.84 L/(m².h bar) and 97%, respectively. This is a remarkable outcome for desalinating water, especially at a pressure of 25 MPa. The fluorinated GNS membrane desalinates water more effec-tively than the hydrogenated GNS membrane at low applied pressures. Although the fluorinated GNS membrane has higher water permeability, it has lower salt rejection at high pressures. A method called molecular dynamics simulation can be employed to predict water desalination performance to a significant extent (Jafarzadeh et al., 2018).

Due to the presence of hydrophilic groups on their surface, F-NPGO films have a substantially larger hydrophilic capacity than F- NPG membranes. Molecular dynamics simulation was used to examine a maximum water permeability value of 3345 L/(m².h bar) at 150 MPa pressure and pore diameters 0.3–0.45 nm for use in water desalination. In comparison to the F-NPG membrane, the hydrophilic functional groups on the surface of the F-NPGO membrane have a higher water permeability.

These functional groups have an impact on the water flow regime. The F-NPGO system has a higher water flow and salt removal rate than the F-NPG system. The advantages of hydroxyl group is increasing water permeability and of hydrogen by simultaneously improving the salt removal rate are both shared by the chemical function of the pores in fluorines (Hosseini et al., 2019).

### 6.2.5 Amino-Functionalized

Magnetic nanoparticles (MNPs) have the ability to remove various aquatic contaminants, such as pathogens, radioactive elements, organic pollutants, and heavy metals, and may be crucial in ensuring that drinking water is safe. To treat Cr(VI) and Ni(II) in wastewater, Norouzian Baghani et al. (2016) fabricated primary amine functional magnetic nano-adsorbents ($Fe_3O_4$-$NH_2$). 1,6-Hexanediamine and $FeCl_3.6H_2O$ were combined to create $Fe_3O_4$-$NH_2$ MNPs in a single step, and after the reaction was complete, the MNPs were simply collected using magnets for additional washing and drying of the end result. At pH 3 and pH 6 at 25°C, the highest adsorption capacities (qm) for Cr(VI) and Ni(II) were 232.51 and 222.12 mg.g$^{-1}$, respectively. Electrostatic attraction and coordination were the main methods used to produce ion adsorption.

At higher pH, amine groups on the adsorbent deprotonated to coordinate with Cr(VI) and Ni(II), but the electrostatic absorbance exceeded the coordination. $Fe_3O_4$-$NH_2$ maintained roughly 76.19% and 77.13% of its original Cr(VI) and Ni(II) adsorption capacity, respectively, after 5 reuses in the presence of 0.2 mol/L NaOH and 0.05 mol/L $HNO_3$. For Cr(VI), $Fe_3O_4$-$NH_2$ demonstrated removal efficiencies of 97.94% and 98.56% from tap water and industrial wastewater, respectively, at an initial concentration of 1.0 mg/L.

A magnetic nanoadsorbent with an amino content of 1.64 mmol/g, an average diameter of 17.9 nm, and a core–shell structure was developed for adsorptive removal of tannic acid (TA) in solution. The results demonstrated the remarkable ability of $Fe_3O_4@SiO_2$-$NH_2$ nanoparticles to adsorb TA in solution. The TA adsorption kinetics fit well with pseudo-second-order kinetics. Ionic strength and solution pH, two elements that affect the chemistry of water, have a huge impact on adsorption capacity. After three desorption–regeneration cycles, the TA adsorption amount decreased from 102.37 to 91.90 mg/g, and no significant drop was observed. Adsorption quantity during the fourth desorption–regeneration cycle remained at 58.64 mg/g, demonstrating that the regenerated adsorbent still had a significant capacity for adsorption and could be applied again to remove TA (J. Wang et al., 2011).

### 6.2.6 Functionalized CNTs

For applications involving the desalination and purification of water, CNTs can generate durable pores in membranes, enhance their antifouling, self-cleaning, and reusable properties (Das et al., 2014). Large surface area, broad porosity ranges, excellent thermal and electrical conductivity, mechanical strength and durability, and high conductivity are the characteristics of CNTs and graphene. They are anticipated to have significant salt rejection and water permeability. Zeolites have three-dimensional networks that can successfully reject salt ions; however, they have low water permeability (Y. Wang et al., 2017).

CNT-ODA and CNT-COOH dispersions were incorporated into the feed and permeate sides of a bilayer polytetrafluoroethylene (PTFE) membrane to enhance the flow in direct contact membrane distillation (DCMD). The desalination efficiency was assessed using the pristine membrane and a monolayer membrane with CNT-ODA on the feed side. The flux through the bilayer membrane increased steadily and reached 126 L/m².h. The bilayer membrane has the highest mass transfer coefficients of all the membranes. Despite being used for extended periods of time, it remained stable. According to the evidence provided here, it is obvious that membrane performance can be improved by including two distinct functionalized CNTs on the two membrane surfaces (Bhadra et al., 2016). The results of these studies show that for both the plain PA membrane and the nanocomposite membrane containing 20% CNTs, the salt rejection ratio increases with increasing ion concentration.

Although Fornasiero et al. (2008) found the opposite effect, in which the ion rejection ratio decreased with increasing ion concentration in the feed and dropped to zero when the ion concentration was equal to 10 mM for KCl, the increase in rejection ratio with increasing ion concentration is an unexpected result. The PA membrane could accommodate the zwitterion-functionalized CNTs, and as the CNT fraction increased, the performance of the composite film increased. The ion removal rate increased or remained constant while the water flux increased noticeably, indicating that the presence of functionalized CNTs may have activated a complement transport mechanism rather than extending nonspecific membrane holes (Chan et al., 2013).

## 6.2.7 Metal–Organic Frameworks

A MOF is formed by the combination of a metal complex and an organic ligand. They have many interactions between metals and organic ligands, large nanomaterial surface area, low density, and a long-modifiable and modified porous framework (Wei et al., 2021). Compared to conventional inorganic debris with firm frameworks, which typically cause membrane blockage and separation, integrated membranes exhibit improved material compatibility. The ability of MOFs to balance water is one of their key qualities for application in water remedies. Reviews of the water balance ability of MOFs have been a focus since 2012 because this is a factor that could limit their ability to be used in water/wastewater utilities, for example, MOF-5, which is composed of $Zn4O(-COO)_6$ SBU and benzene-1,4-dicarboxylate linkers, and HKUST-1, which is composed of $Cu(-COO)_4$ SBU and 1,3,5-benzene tricarboxylic acid. However, these materials have been reported to be surprisingly porous and thermally strong, which limit their use in water-related treatments.

According to Wang et al., there are only a few methods to develop water-resistant MOFs: (1) using high-valence metal ions and (2) using an azolate-based all-natural linker (Abdullah et al., 2021). Ferey and his colleagues (2005) developed the well-known Fe- and Cr-based MIL-100 and MIL-101, which can offer good chemical stability and persist for many months in ambient and other solvents. Additionally, even in acidic and basic environments, MOFs containing high-valence $Zr^{4+}$ cations, including the well-known UiO-66 and PCN families, demonstrate outstanding hydropower stability (C. Wang et al., 2016).

Numerous case studies on promoting the hydrothermal stability of MOFs have been reported, including (1) Taylor et al.'s finding that the nonpolar alkyl functional groups in CALF-25 allow the structure to absorb a lot of water while maintaining structural stability because the functional groups surrounding the metal core are shielded. (2) A group of fluorinated MOFs (f-MOFs), which are incredibly hydrophobic and have remarkable water stability, have been developed by Nijem et al. (2013). (3) Post-synthesis techniques, such as ligand modification and metal and ligand exchangers, have been devised to considerably increase the hydrophobicity and hydrothermal stability of the already existing MOF structures (C. Wang et al., 2016).

### 6.2.7.1 UiO-66(Zr)-

UiO-66(Zr) is a zirconium-based MOF material composed of 1,4-benzodicarboxylic acid organisms and metal clusters of $[Zr_6O_4 (OH)_4]$. It has excellent chemical stability. The value of aperture is approximately 6.0105 Å. The TFN-UiO (0.1 wt% UiO-66) membrane has a 52% increase in water permeability when compared to an unmodified membrane, and a 95% salt rejection rate. $MgCl_2$, $CaCl_2$, and $Na_2SO_4$ had rejection rates of 97.81%, 92.81%, and 99.23%, respectively. The exceptional ability of imidazole-2-carboxaldehyde (ICA) to bind and speed up the movement of water molecules was found. The internal polarity of the MOF nanofillers is strengthened by the insertion of hydrophilic groups $-NH/NH_2$. This membrane achieves a $Na_2SO_4$ retention rate (97.4%) and is twice as permeable to water as the pristine membrane (Wei et al., 2021).

*Effect of UiO-66 nanoparticles loading:* For TFN-U4 (0.2 wt% loading), TFN-U3 (0.15 wt% loading), TFN-U2 (0.1 wt% loading), TFN-U1 (0.05 wt% loading), and TFC (0 wt% loading), different thicknesses have been found. The findings of the contact angle experiments demonstrate that the layer hydrophilicity declines with increasing UiO-66 nanoparticle content in the range of 0–0.2 wt%; it lowers from 47.8° for pure TFC membrane to 29.6° for TFN-U2 (0.1 wt%) and further decreases to 23.9° when the particle loading increases to 0.2 wt% (Ma et al., 2017).

Gu et al. (2020) developed a lysine-functionalized UiO-66 PES membrane for NF applications using a pressure-assisted technique and an IP reaction. They discovered that the hydrophilic Lys@UiO-66 develops an orderly and stable lattice structure, increasing the membrane's surface area while offering paths for more water molecules. When Lys@UiO-66 was loaded at a rate of 17.13 mg/cm², efficiency was at its highest, and water permeability was 55% higher than that of the PA-TFC membrane (18.27 L/m².h bar) (Wei et al., 2021).

Bagherzadeh et al. (2020) used graphene quantum dots (GQDs) to modify UiO-66-NH$_2$. They observed that GQDs were successfully developed with the GQDs@ UiO-66-NH$_2$ @PAFO film and that their presence enhanced water attraction to the MOF's surface and improved the matrix compatibility of the PA layer. GQDs@ UiO-66-NH$_2$ nanoparticles were added, which enhanced the film's capacity to block salt and water vapor. The salt removal rate was 1.5 times as fast as that of the original film while using TFN-0.25 in the FO mode, which included 250 ppm GQDs@ UiO-66-NH$_2$.

### 6.2.7.2 Zeolitic imidazolate frameworks (ZIFs)

MD efficiency can be improved by applying a ZIF-8/chitosan-containing ultra-thin layer to the polyvinylidene fluoride (PVDF) membrane. The air gap membrane distillation (AGMD) experiment found that water permeability (7.2 L/m².h) was improved by 3.5 times compared to that of the unmodified membrane. The NaCl waterproofing rate was up to 99.5%, and the efficiency was increased. It has significantly better antifouling than the unaltered film, and its flux recovery is 90% (>67%). In order to develop the ZIF-67/PPy CDI electrode, Wei et al. (2021) coupled ZIF-67 with regular micropores and high porosity with PPy nanotubes with high conductivity. With a desalination capability of 11.34 mg/g, it can effectively purify salt water.

## 6.2.8 Covalent Organic Frameworks

Compared with inorganic nanomaterials, the organic covalent bonds of covalent organic frameworks (COFs) have better affinity for polymers. Their excellent thermal and chemical stability gives them excellent stability. These membranes will become a new generation of nanomaterials for seawater desalination.

Xiao et al. (2021) fabricated the bi-layer COF nanofilm. They developed a composite layer before the COF on top of another so that the films had a clear, flawless multilayer structure. It should be noted that they adapted a two-dimensional COF multilayer structure on the membrane interface to prepare narrow channels. The main reason was that the two nanofilms migrated at the interface during the second growth, resulting in a narrow pore size distribution and better ion selectivity. The removal rate of Na$_2$SO$_4$ was 95.8%. This work will promote the growth of COF-based membranes with sub-nano channels. It can also be applied to other frame-based materials.

Wu et al. (2019) conducted an experiment with a highly efficient NF composite film by combining dopamine with a COF after interfacial polymerization. The COF interlayer controls the adsorption/diffusion of the amine monomer during interfacial polymerization. In addition, the interlayer also improves the interface interaction between the PA layer and the PAN support layer, showing excellent hydrophilicity and structural stability. The PA/PDA-COF/PAN NF membrane has a high desalination rate (the removal rate of Na$_2$SO$_4$ was 93.4%) and dye removal (orange GII was 94.5%), and the permeability flux was 207.07 L/m² h MPa.

# References

Abd Aziz, M. H., Dzarfan Othman, M. H., Alias, N. H., Nakayama, T., Shingaya, Y., Hashim, N. A., Kurniawan, T. A., Matsuura, T., Rahman, M. A., & Jaafar, J. (2020). Enhanced omniphobicity of mullite hollow fiber membrane with organosilane-functionalized TiO$_2$ micro-flowers and nanorods layer deposition for desalination using direct contact membrane distillation. *Journal of Membrane Science, 607*, 118137. https://doi.org/10.1016/j.memsci.2020.118137

Abdelsamad, A. M. A., Khalil, A. S. G., & Ulbricht, M. (2018). Influence of controlled functionalization of mesoporous silica nanoparticles as tailored fillers for thin-film nanocomposite membranes on desalination performance. *Journal of Membrane Science, 563*, 149–161. https://doi.org/10.1016/j.memsci.2018.05.043

Abdullah, N., Yusof, N., Ismail, A. F., & Lau, W. J. (2021). Insights into metal-organic frameworks-integrated membranes for desalination process: A review. *Desalination, 500*, 114867. https://doi.org/10.1016/j.desal.2020.114867

Al-Karaghouli, A., Renne, D., & Kazmerski, L. L. (2009). Solar and wind opportunities for water desalination in the Arab regions. *Renewable and Sustainable Energy Reviews, 13*(9), 2397–2407. https://doi.org/10.1016/j.rser.2008.05.007

Baghani, N. A., Mahvi, A. H., Gholami, M., Rastkari, N., & Delikhoon, M. (2016). One-Pot synthesis, characterization and adsorption studies of amine-functionalized magnetite nanoparticles for removal of Cr (VI) and Ni (II) ions from aqueous solution: Kinetic, isotherm and thermodynamic studies. *Journal of Environmental Health Science and Engineering, 14*, 1–2. https://doi.org/10.1186/s40201-016-0252-0

Bagherzadeh, M., Bayrami, A., & Amini, M. (2020). Enhancing forward osmosis (FO) performance of polyethersulfone/polyamide (PES/PA) thin-film composite membrane via the incorporation of GQDs@UiO-66-NH2 particles. *Journal of Water Process Engineering, 33*, 101107. https://doi.org/10.1016/j.jwpe.2019.101107

Baig, M. I., Ingole, P. G., Jeon, J., Hong, S. U., Choi, W. K., Jang, B., & Lee, H. K. (2019). Water vapor selective thin film nanocomposite membranes prepared by functionalized silicon nanoparticles. *Desalination, 451*, 59–71. https://doi.org/10.1016/j.desal.2017.06.005

Bhadra, M., Roy, S., & Mitra, S. (2016). A bilayered structure comprised of functionalized carbon nanotubes for desalination by membrane distillation. *ACS Applied Materials & Interfaces, 8*(30), 19507–19513. https://doi.org/10.1021/acsami.6b05644

Chan, W.-F., Chen, H., Surapathi, A., Taylor, M. G., Shao, X., Marand, E., & Johnson, J. K. (2013). Zwitterion functionalized carbon nanotube/polyamide nanocomposite membranes for water desalination. *ACS Nano, 7*(6), 5308–5319. https://doi.org/10.1021/nn4011494

Das, R., Ali, Md. E., Hamid, S. B. A., Ramakrishna, S., & Chowdhury, Z. Z. (2014). Carbon nanotube membranes for water purification: A bright future in water desalination. *Desalination, 336*, 97–109. https://doi.org/10.1016/j.desal.2013.12.026

Férey, G., Mellot-Draznieks, C., Serre, C., Millange, F., Dutour, J., Surblé, S., & Margiolaki, I. (2005). A chromium terephthalate-based solid with unusually large pore volumes and surface area. *Science, 309*(5743), 2040–2042. https://doi.org/10.1126/science.1116275

Fornasiero, F., Park, H. G., Holt, J. K., Stadermann, M., Grigoropoulos, C. P., Noy, A., & Bakajin, O. (2008). Ion exclusion by sub-2-nm carbon nanotube pores. *Proceedings of the National Academy of Sciences, 105*(45), 17250–17255. https://doi.org/10.1073/pnas.0710437105

Gu, Z., Yu, S., Zhu, J., Li, P., Gao, X., & Zhang, R. (2020). Incorporation of lysine-modified UiO-66 for the construction of thin-film nanocomposite nanofiltration membrane with enhanced water flux and salt selectivity. *Desalination, 493*, 114661. https://doi. org/10.1016/j.desal.2020.114661

Hosseini, M., Azamat, J., & Erfan-Niya, H. (2019). Water desalination through fluorine-functionalized nanoporous graphene oxide membranes. *Materials Chemistry and Physics, 223*, 277–286. https://doi.org/10.1016/j.matchemphys.2018.10.063

Huang, J., Li, Q., Sun, D., Lu, Y., Su, Y., Yang, X., Wang, H., Wang, Y., Shao, W., He, N., Hong, J., & Chen, C. (2007). Biosynthesis of silver and gold nanoparticles by novel sundried *Cinnamomum camphora* leaf. *Nanotechnology, 18*(10), 105104. https://doi. org/10.1088/0957-4484/18/10/105104

Jafarzadeh, R., Azamat, J., & Erfan-Niya, H. (2018). Fluorine-functionalized nanoporous graphene as an effective membrane for water desalination. *Structural Chemistry, 29*(6), 1845–1852. https://doi.org/10.1007/s11224-018-1162-9

Kotp, Y. H. (2021). High-flux TFN nanofiltration membranes incorporated with Camphor-Al$_2$O$_3$ nanoparticles for brackish water desalination. *Chemosphere, 265*, 128999. https:// doi.org/10.1016/j.chemosphere.2020.128999

Ma, D., Peh, S. B., Han, G., & Chen, S. B. (2017). Thin-film nanocomposite (TFN) membranes incorporated with super-hydrophilic metal—Organic framework (MOF) UiO-66: Toward enhancement of water flux and salt rejection. *ACS Applied Materials & Interfaces, 9*(8), 7523–7534. https://doi.org/10.1021/acsami.6b14223

Nijem, N., Canepa, P., Kaipa, U., Tan, K., Roodenko, K., Tekarli, S., Halbert, J., Oswald, I. W. H., Arvapally, R. K., Yang, C., Thonhauser, T., Omary, M. A., & Chabal, Y. J. (2013). Water cluster confinement and methane adsorption in the hydrophobic cavities of a fluorinated metal—Organic framework. *Journal of the American Chemical Society, 135*(34), 12615–12626. https://doi.org/10.1021/ja400754p

Pang, R., & Zhang, K. (2018). Fabrication of hydrophobic fluorinated silica-polyamide thin film nanocomposite reverse osmosis membranes with dramatically improved salt rejection. *Journal of Colloid and Interface Science, 510*, 127–132. https://doi.org/10.1016/j. jcis.2017.09.062

Siddeeg, S. M., Tahoon, M. A., Alsaiari, N. S., Shabbir, M., & Rebah, F. B. (2021). Application of functionalized nanomaterials as effective adsorbents for the removal of heavy metals from wastewater: A review. *Current Analytical Chemistry, 17*(1), 4–22. https://doi.org/1 0.2174/1573411016999200719231712

Teow, Y. H., & Mohammad, A. W. (2019). New generation nanomaterials for water desalination: A review. *Desalination, 451*, 2–17. https://doi.org/10.1016/j.desal.2017.11.041

Tiraferri, A., Kang, Y., Giannelis, E. P., & Elimelech, M. (2012). Highly hydrophilic thin-film composite forward osmosis membranes functionalized with surface-tailored nanoparticles. *ACS Applied Materials & Interfaces, 4*(9), 5044–5053. https://doi.org/10.1021/ am301532g

Wang, C., Liu, X., Demir, N. K., Paul Chen, J., & Li, K. (2016). Applications of water stable metal—organic frameworks. *Chemical Society Reviews, 45*(18), 5107–5134. https://doi. org/10.1039/C6CS00362A

Wang, J., Zheng, C., Ding, S., Ma, H., & Ji, Y. (2011). Behaviors and mechanisms of tannic acid adsorption on an amino-functionalized magnetic nanoadsorbent. *Desalination, 273*(2), 285–291. https://doi.org/10.1016/j.desal.2011.01.042

Wang, X., Wang, H., Wang, Y., Gao, J., Liu, J., & Zhang, Y. (2019). Hydrotalcite/graphene oxide hybrid nanosheets functionalized nanofiltration membrane for desalination. *Desalination, 451*, 209–218. https://doi.org/10.1016/j.desal.2017.05.012

Wang, Y., He, Z., Gupta, K. M., Shi, Q., & Lu, R. (2017). Molecular dynamics study on water desalination through functionalized nanoporous graphene. *Carbon, 116*, 120–127. https://doi.org/10.1016/j.carbon.2017.01.099

Wei, H., Zhao, S., Zhang, X., Wen, B., & Su, Z. (2021). The future of freshwater access: Functional material-based nano-membranes for desalination. *Materials Today Energy, 22*, 100856. https://doi.org/10.1016/j.mtener.2021.100856

Wu, M., Yuan, J., Wu, H., Su, Y., Yang, H., You, X., Zhang, R., He, X., Khan, N. A., Kasher, R., & Jiang, Z. (2019). Ultrathin nanofiltration membrane with polydopamine-covalent organic framework interlayer for enhanced permeability and structural stability. *Journal of Membrane Science, 576*, 131–141. https://doi.org/10.1016/j.memsci.2019.01.040

Xiao, A., Shi, X., Zhang, Z., Yin, C., Xiong, S., & Wang, Y. (2021). Secondary growth of bi-layered covalent organic framework nanofilms with offset channels for desalination. *Journal of Membrane Science, 624*, 119122. https://doi.org/10.1016/j.memsci.2021.119122

Yang, X., Li, Q., Wang, H., Huang, J., Lin, L., Wang, W., Sun, D., Su, Y., Opiyo, J. B., Hong, L., Wang, Y., He, N., & Jia, L. (2010). Green synthesis of palladium nanoparticles using broth of *Cinnamomum camphora* leaf. *Journal of Nanoparticle Research, 12*(5), 1589–1598. https://doi.org/10.1007/s11051-009-9675-1

Zargar, M., Hartanto, Y., Jin, B., & Dai, S. (2017). Understanding functionalized silica nanoparticles incorporation in thin film composite membranes: Interactions and desalination performance. *Journal of Membrane Science, 521*, 53–64. https://doi.org/10.1016/j.memsci.2016.08.069

# 7

# Functional Nanomaterials and Polymeric Materials for Bioremediation

## 7.1 Overview of Bioremediation Process

Environmental pollution is a very serious issue today, as contamination from hazardous waste has led to low availability of clean water and soil disturbances which have hampered the production of crops. Bioremediation of contaminated water involves the use of biological agents, especially microorganisms such as bacteria, fungi, and yeast. Microorganisms use contaminants as food or energy sources in bioremediation processes. In order for bioremediation to take place, three key components are required: microorganisms, food, and nutrients. The bioremediation triangle refers to these three main components (Kulshreshtha et al., 2014).

Depending on the nature of their activities, there are two types of bioremediation: *in situ* bioremediation, in which the contaminants are treated in a common location, and *ex situ* bioremediation, whereby the contaminants are removed at different locations (Rando et al., 2022). Heavy metals, e.g., $Cd^{2+}$, $Pb^{2+}$, or $Mn^{2+}$, are highly toxic and pose a significant risk to the environment due to their nondegradable nature. Heavy metals are elements with a density of more than 5.0 g $cm^{3.}$ (Kaur & Roy, 2021). They are also sometimes called transition and post-transition metals (Hawkes, 1997).

Heavy metals are not biodegradable and, therefore, have an impact on both local flora and fauna and human health. The problem of removing large amounts of metal poses a serious challenge, and traditional approaches have their drawbacks. Therefore, alternative means of effectively removing heavy metals need to be found (Kaur & Roy, 2021). Nanotechnology and novel nanomaterials are currently employed in a wide range of applications due to features such as size effect, catalytic capacity, large surface area (surface area/volume ratio), and reactivity (Rando et al., 2022).

Nanoparticles are classified as organic nanoparticles (micelles, fullerenes, and dendrimers) and inorganic nanoparticles (ceramic, steel, and metal oxide nanoparticles) (Sharma & Sharma, 2022) (Figure 7.1). So Heavy metal pollution can be reduced with the help of nanomaterials manufactured using environmentally friendly processes (Malik et al., 2022). In particular, carbon nanotubes (CNTs), graphene, and their derivatives have attracted the scientific community for biological remediation due to their large specific surface area and concentrated pore size distribution (Sudhakar et al., 2020).

DOI: 10.1201/9781003391364-7

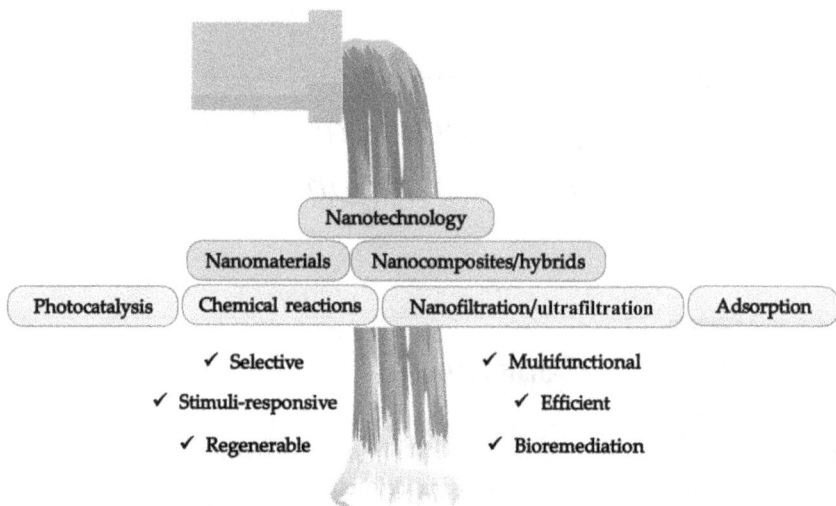

**FIGURE 7.1**

Various approaches based on nanotechnology for the removal of environmental pollutants and bioremediation (Rando et al., 2022). *Reproduced with permission from MDPI.*

Apart from heavy metals, there is oil/water separation. Crude oils can be in the form of water-in-oil (w/o) or oil-in-water (o/w) emulsions. The type of emulsion is determined by the ratio of oil and water produced. Every barrel of oil transported around the world by waterways endangers the environment through spills. Oil spills have been occurring for decades, and unfortunately, cleanup techniques have changed very little over the time.

A number of materials can be used as a base for nanoparticle synthesis, including iron oxides $Fe_2O_3/Fe_3O_4$, pure metals Fe and Co, and spinel-type ferromagnets $MgFe_2O_4$, $MnFe_2O_4$, or $CoFe_2O_4$ (Simonsen et al., 2018).

In particular, nanocomposites and nano-hybrids are considered based on the following:

- Metal oxide nanoparticles
- Metal nanoparticles
- Carbon-based nanomaterials

## 7.2 Nanoadsorbents

Nanoadsorbents of metal oxides, including iron oxide, titanium dioxide ($TiO_2$), and nickel oxide, are affordable and have strong adsorption capacity, which make them easy to recycle. Their capacity to adsorb particles is dependent on their size. Their adsorption capacity enhanced as the particle size decreased.

## 7.3 Heavy Metal Separation Techniques

Recent studies have explored the use of materials such as CNTs, nanocomposites, nanospheres, nanofibers, nanoclay, and nanowires in combination with conventional wastewater treatment techniques that may aid in the removal of various organic and inorganic contaminants, including heavy metals. The potential for use has been highlighted (Kaur & Roy, 2021).

Various techniques have been employed for the synthesis of nanoparticles for bottom-up approaches, such as laser pyrolysis, sol-gel, aerosol-based processes, plasma spray processes, and green synthesis (Sharma & Sharma, 2022). For instance, modification of a polymeric membrane using clay nanoparticles is illustrated in Figure 7.2.

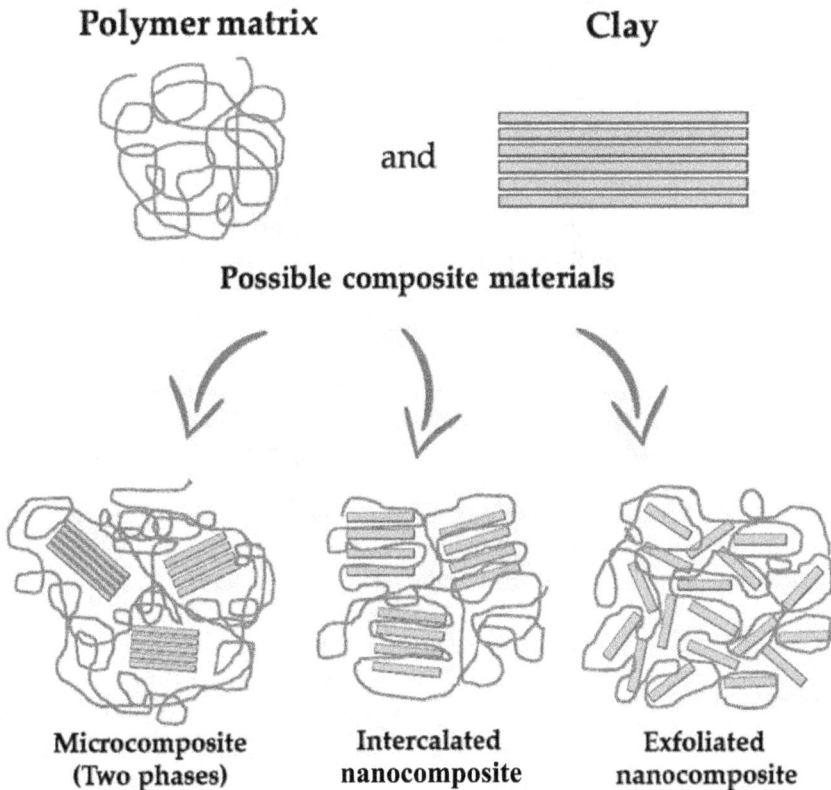

**FIGURE 7.2**
Functionalization of polymeric membrane matrix with clay nanoparticles. The possible method of forming nanocomposites is: (i) micro composite (Two-phases), (ii) intercalated nanocomposites, (iii) Exfoliated nanocomposites.

### 7.3.1 Metal Nanoparticles

Metal nanoparticles (MNPs; M = related to noble metals or metals such as Ni, Al, Au, Pt, Ru, Ag, Pd, or Cu) are physically and chemically different from bulk materials due to their small size and weight. MNPs are nanomaterials with unique properties, such as surface-to-surface volume ratio. Surface functionalization of those nanoparticles with various functional groups and modification of polymeric membranes with functionalized nanomaterials help enhance wastewater treatment in an effective way.

Poly(vinylidene fluoride) (PVDF) membranes grafted with nanocomposites of AgNPs-PCBMA (silver nanoparticles and poly(carboxybetaine methacrylate)) have demonstrated anti-protein and antimicrobial fouling properties. Polypropylene hollow fiber membranes functionalized with osmium NPs were prepared, obtained *in situ* from a solution of osmium tetroxide in tert-butyl alcohol by reduction with molecular hydrogen, and placed in a membrane contactor.

Oxidation of environmentally harmful organic compounds such as methylene blue is possible using olfactory photo catalysts in water. For example, a Pd-AgNPs/macroporous silicon (macroPSi) bimetallic plasmonic photocatalyst-based heterostructure might be used to enhance the activity of methylene blue decomposition in water under UV irradiation.

This heterostructure is created by dipping macroPSi in a simple technique that deposits monometallic and bimetallic Ag and Pd NPs. Although monometallic photocatalysts AgNPs/macroPSi and PdNPs/macroPSi may be made, the bimetallic photocatalyst performs better with greater efficiency (98.8%) and methylene blue rejection rate (0.033 min$^{-1}$) due to the largest specific surface area and plasmonic effect. As visible-light selective photocatalysts in water, several Au/Bi$_2$WO$_6$ nanocomposites can be developed via a hydrothermal process combined with a fast reduction–deposition technique and varied weight ratios of Au. In particular, under visible light and aerobic circumstances, this hybrid nanostructure demonstrates a significant capacity for benzylic alcohol oxidation and Cr(VI) reduction in water. Moreover, 2.0 wt% and 1.0 wt% Au/Bi$_2$WO$_6$ are the best catalysts for this oxidation/reduction process, respectively (Wali et al., 2019).

Composites based on cellulose and AgNPs can be easily made. Composite nanomaterials with antimicrobial, antioxidant, and photodegradable properties were developed simply by impregnating cellulose derived from citrus waste with AgNPs (Rando et al., 2022). Ali et al. (2017) studied the antibacterial ability of AgNP-based modified cellulose membrane by preparing discs of cellulose-AgNP composites, which showed >90% reduction in *Staphylococcus aureus* culture within 150 min and moderate overall antioxidant capacity, and 2,2-diphenyl-1-picryl-hydrazil (DPPH) has low radical scavenging activity and moderate photolysis. Methylene blue dye's capacity in sunlight is up to 63.16% (60 minutes).

Mesoporous Fe$_3$O$_4$ nanoparticles with amine functionality (AF-Fe$_3$O$_4$) were synthesized to remove hazardous heavy metal ions from water with extreme efficiency. A novel ecologically friendly and cost-effective manufacturing procedure was used to produce AF-Fe$_3$O$_4$. The largest amino group that can be grafted onto AF-Fe$_3$O$_4$ is 0.1790 g/mg, according to the ninhydrin test. By using low-field magnetic separation, it is simple to recover AF-Fe$_3$O$_4$ from water in less than a minute. At pH 7.0, equilibrium was attained after 120 minutes of the adsorption of 5 mg/l Pb(II), Cd(II), and Cu(II) on 50 ml of 10 mg AF-Fe$_3$O$_4$. The maximum adsorption capacities for Pb(II), Cd(II), and Cu(II), which are greater than prior findings, range from 369.0 to 523.6 mg/g and are in good agreement with the Langmuir adsorption isotherm model. In particular, the adsorption of heavy metal ions onto mesoporous magnetic nanoparticles is due to the interesting properties of heavy metal ion adsorption such as surface functionalization (Xin et al., 2012). According to the ninhydrin assay, AF-MSNs are uniformly sized around 100 nm and loaded with 1.2 mg/g of amino groups. The absorption of metal and metal oxide nanoparticles and photocatalysis-based degradation are demonstrated in Figure 7.3.

For mesoporous carbon, the cumulative Barrett-Joyner-Halenda (BJH) desorption volume of the pores is 0.1279 cm$^3$/g, with an average pore diameter of 3.84 nm. Fe$_3$O$_4$ has a size of roughly 20 nm on average. Hydrothermal process was used to develop AF-Fe$_3$O$_4$ nanoparticles. A transparent solution of FeCl$_3$ 6H$_2$O (1 g) was commonly formed by dissolving it in ethylene glycol (20 mL), adding sodium acetate (NaAc) (3 g), and then adding ethane diamine (10 mL). The liquid was aggressively agitated for 30 minutes before being vacuum-sealed in a stainless steel autoclave with Teflon lining. After being heated and kept at 200°C for 8 hours, the autoclave was allowed to cool to ambient temperature. Water was used to wash the product multiple times. It was possible to use the resulting black precipitate (AF-Fe$_3$O$_4$). The amino group grafted on AF-Fe$_3$O$_4$ was 0.1790 µg/mg according to the ninhydrin test.

In a traditional adsorption procedure, 10 mg of the newly created AF-Fe$_3$O$_4$ was added to 50 ml of a mixed solution comprising 5 mg/l of each heavy metal ion, the

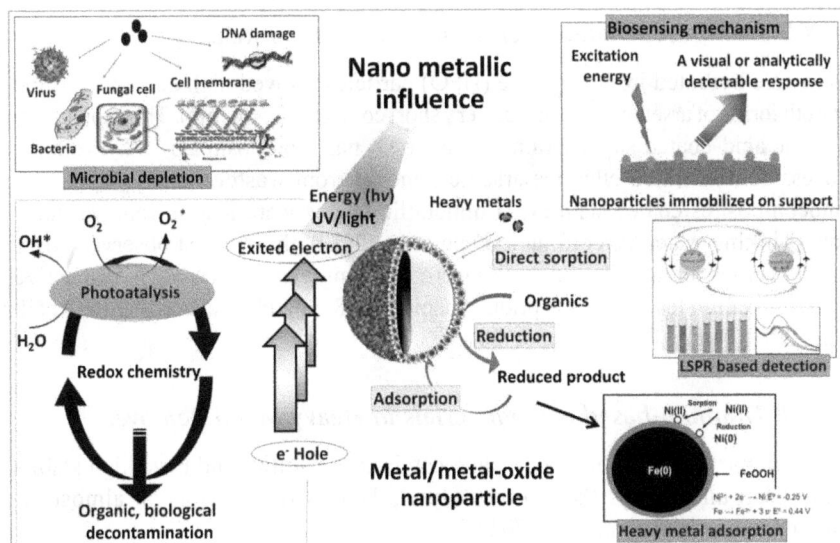

**FIGURE 7.3**
Functionalized metal- and metal oxide-based nanoparticles for bioremediation and degradation of pollutants (Chauhan et al., 2022). *Reproduced with permission from Elsevier.*

pH of which was then raised to 7.0 by adding HCl and NaOH, and the mixture was stirred for hours. After 120 min, AF-Fe$_3$O$_4$ with adsorbed heavy metal ions was separated from the mixture with a hand permanent magnet. Residual heavy metals in solution were determined by ICP-MS. At temperatures ranging from 293 to 313 K, adsorption onto AF-Fe$_3$O$_4$ was performed with various concentrations of each metal ion (Xin et al., 2012; J. Wang et al., 2010).

Metal oxides such as iron oxide, TiO$_2$, and nickel oxide are inexpensive, have high adsorption capacity, and are easy to regenerate (Kaur & Roy, 2021).

### 7.3.1.1 Role of Silver (Ag) Nanoparticles in Heavy Metal Removal

Mercury ions were shown to be more effectively taken up by silver nanoparticles. We found that the mesoporous nano silica employed to remove mercury from wastewater may efficiently absorb mercury ions. *Ficus benjamina* leaf extract was used to create zero-valent Ag nanoparticles to successfully remove cadmium (Al-Qahtani, 2017). Adsorption onto AF-Fe$_3$O$_4$ was carried out at temperatures ranging from 293 to 313 K with different initial concentrations of individual metal ions.

Both planktonic cells and biofilms of Gram-negative *Escherichia coli* and *Pseudomonas aeruginosa* were resistant to Gram-negative hybrid nanosilica made with silver nanoparticles (NSAgNPs) (K. Das et al., 2013). Similarly, Shi et al. (2011) examined the removal efficiency of mercury in natural water using AgNP-immobilized nanofibers. From there, they concluded that nylon membrane filters loaded with silver nanoparticles displayed remarkable efficiency of mercury accumulation from the aqueous phase.

### 7.3.1.2 Iron-based Nanomaterials in Heavy Metal Removal

Nanoscale hydrated iron(III) oxide (HFO) particles showed high adsorption affinity for both forms of arsenic and required very short contact times (4 min). The application of ascorbic acid-coated superparamagnetic Fe$_3$O$_4$ nanocomposites by a hydrothermal process demonstrated effective arsenic removal from wastewater. Fe$_3$O$_4$ magnetic nanocomposites coated with Fe-Ti bimetallic oxides were able to remove fluoride from drinking water. According to Feng et al. (2012), the greatest observed adsorption capacities for As(III) and As(V) were 16.56 mg/g and 46.06 mg/g, respectively. Fe$_3$O$_4$ has a high lead ion adsorption capability of 83 mg Pb/g, according to (Recillas et al., 2011; Cumbal & SenGupta, 2005).

### 7.3.1.3 Titanium-based Nanomaterials in Heavy Metal Removal

TiO$_2$ nanoparticles exhibit strong photocatalytic capability and have high stability, cheap cost, and safety. For photocatalysis, TiO$_2$ is regarded as an almost ideal semiconductor (Chauhan et al., 2022).

When comparing partition coefficients, TiO$_2$ nanoparticles performed better than other metal oxide nanoparticles. TiO$_2$ nanoparticles with sizes ranging from 7 to 10 nm and specific surface areas ranging from 9.5 to 185.5 m²/g (calculated by BET measurements) have demonstrated the ability to oxidize Pb, Cd, Zn, Ni, and Cu impurities (Dunphy Guzman et al., 2006).

Recillas et al. (2011) carried out an experimental investigation using anatase nanoadsorbents made of titanium to remove lead, copper, and arsenic. From there, they noticed that $TiO_2$ has an adsorption capability of 159 mg/g, which allows it to remove lead ions. In a study by Kaur and Roy (2021), titanium adsorbents were found to have the highest capacity for adsorbing lead (31.25 mg/g), copper (23.74 mg/g), and arsenic (16.98 mg/g), with the capacity for lead and copper-lead increasing with increasing pH.

### 7.3.1.4 Manganese-based Nanomaterials in Heavy Metal Removal

Fe-MN binary oxide nanomaterial is effective in removing both As(V) and As(III), and it entirely oxidized As(III) to As(V). Manganese-iron oxide (MnFe2O4) nanoparticles have reported maximum chromium adsorption and need extremely limited contact time.

The Fe-Mn binary oxide has twice the adsorption capacity compared to $MnFe_2O_4$ and $CoFe_2O_4$ due to the higher number of surface OH groups. Also, 0.1M NaOH can regenerate 80–90% of this nanomaterial within 24 hours of contact time. The long contact time required for regeneration was one of its drawbacks. MnO has also been proven in removing arsenic from wastewater (S. Zhang et al., 2010, p. 4).

### 7.3.1.5 Carbon-based Nanomaterials in Heavy Metal Removal

Multi-walled carbon nanotubes (MWCNTs) result in enhanced adsorption efficiency of Cd (II) and, hence, are suggested for use in treating cadmium-contaminated water (Bhanjana et al., 2017). Chromium(VI), arsenic, and copper–ammonia complexes were successfully removed from water using hybrid metal oxide nanoparticles that incorporate MWCNTs. It was possible to develop magnetic iron oxide nanoparticles that had been 2-mercaptoben-zothiazole modified in order to quickly and effectively remove mercury from wastewater. With a rejection rate of over 90%, magnetic nanoparticles modified with cetyltrimethylammonium bromide (CTAB) demonstrated high adsorption capacity for arsenate.

### 7.3.1.6 Other Inorganic Nanomaterials for Heavy Metal Removal

Biochar can be used in the preparation of Fe/BC composites for the remediation of heavy metals and organic compounds in the environment through the use of pyrolysis and hydrothermal carbonization, ball milling with low energy consumption, and fractional precipitation, in addition to NPs (and occasionally stabilizers like carboxymethyl cellulose) such as Fe NPs for the production of nanocomposites.

Metal oxide NPs like $TiO_2$, ZnO, and $Bi_2O_3$ have the ability to photocatalyze the degradation of organic dyes like methylene blue and other dyes.

An ultrasonic-assisted sol-gel procedure was used to produce magnetic $TiO_2$ that was then hybridized with lignocellulosic biomass (olive pits, OP) and hydrothermally magnetized at 180°C to form $TiO_2$-OP@$Fe_3O_4$. Rhodamine B, methylene blue, Congo red, and other dyes, as well as Cr(VI), were removed from contaminated aquatic systems through absorption and photocatalytic processes when exposed to visible light (Djellabi et al., 2019).

## 7.3.2 Functionalized Polymers and Nanomaterials for Oil–Water Separation

Oil spill cleanup and recovery also uses polymer- and NP-based nanocomposites. The co-assembly of amphiphilic block copolymers of poly(acrylic acid)-block-polystyrene (PAA20-b-PS280) and oleic acid-stabilized magnetic iron oxide NPs results in magnetic shell cross-linked knedel-like nanoparticles (MSCKs) with amphiphilic organic domains and a magnetic sensitive core that are used for the adsorption of hydrophobic guest molecules.

Halloysite nanotubes (HNTs) are made of halloysite and loaded with surfactants. HNTs have the ability to both create stable pickering emulsions and load surfactants for targeted release at the oil–water interface. At the oil–water interface, these dispersants can be released selectively, which can drastically cut down on the amount of surfactants needed. A wax covering that dissolves when in contact with oil was produced in a recent study as a means of controlling the release of surfactants from HNTs.

Pickering emulsions is stabilized by solid particles, that adhere to the oil–water interface and stop droplet coalescence. The monolayer that results from irreversible particle adsorption at the oil–water interface prevents coalescence and creates a stable emulsion. The wettability of the particles has a significant impact on the stability of pickering oil-in-water emulsions. To avoid the particles remaining in the pure aqueous phase, the particles should have a contact angle of > 90° and be somewhat hydrophobic (Pete et al., 2021).

At 5000 ppm, the bionanocompounds magnetite NP-PEA-OmpA and magnetite NPPEA-OmpA-laccase successfully separated a 1 wt% oil/water emulsion while also demonstrating heavy crude oil removal efficiency of 81% and 88%, respectively, and degradation levels between 5% and 50% (Rangel-Muñoz et al., 2020).

### 7.3.2.1 Carbon-based Materials for Removal of Dyes and Oil/Water

Due to the way that hybrid polyurethane (PU) materials combine the superior absorption capabilities of PU with the surface characteristics of nanomaterials, they have received a lot of interest as oil absorbers in oil–water separation. A two-step procedure was devised for incorporating CNTs into a porous PU foam framework to develop hybrid materials that absorb oil. MWCNTs were first altered by oxidation in an $H_2SO_4$–$HNO_3$ binary mixed acid, and then a silane coupling agent (KH 570) was grafted onto their surface. Then, employing isocyanate and polyether polyol as monomers for polymerization and surface-modified MWCNTs as inorganic components, MWCNT/PU hybrid materials were created. The surface-grafted MWCNTs exhibit extremely low water contact angles, up to 153° (T. Zhang et al., 2018).

### 7.3.2.2 Functionalization of Nanodiamond Materials with Octadecylamine for Removal of Oil/Water

The functional modification of nanodiamonds (NDs), promising members of the carbon family, was achieved through the covalent attachment of octadecylamine (ODA). Then, using a rapid dip-drying procedure to embed her ND-ODA into a melamine sponge (MS) scaffold, we created an extremely hydrophobic sponge with a hierarchical microstructure. This technique resulted in an extremely hydrophobic sponge (ND-ODA@PDMS@MS) with a water contact angle of 155 ± 2°.

For different oils and organic solvents, ND-ODA@PDMS@MS exhibits outstanding oil-water separation efficiency (above 98.6%) and absorption capacity (26.65–55.64 g/g). Additionally, the crude oil adsorption capacity is relatively stable in strong acid, alkaline, and salty environments, ensuring its use in the cleanup of maritime oil spills and industrial oil wastes. In addition to one-off oil absorption, ND-ODA@PDMS@MS demonstrated satisfactory performance in continuous oil–water separation.

For effective oil–water separation, super hydrophobic PU was modified by NDs and precoated with polydopamine and 1H,1H,2H,2H-perfluorodecanethiol (PFDT). Due to their innate hydrophobicity, the work focuses on the direct production of highly hydrophobic PDMS/sponge sorbents for oil–water separation. We therefore postulate that the hierarchical structure of the ND-ODA-modified sponge and the hydrophobicity of both ND-ODA and PDMS can result in enhanced performance in oil–water separation. A better super hydrophobic sponge (ND-ODA@PDMS@MS, made using a dip-coating technique) may efficiently separate oil from water by selectively adsorbing oil and repelling water. High hydrophobicity (157°), an adsorption capacity of up to 26.65–55.64 times the original weight, and exceptional mechanical qualities all contribute to the reusability of ND-ODA@PDMS@MS. Gravimetric measurement of various oils and solvents were used to estimate the adsorption capability of ND-ODA@PDMS@MS (H. Wang et al., 2022).

### 7.3.2.3 Commercial Polymeric Spongy Materials for Oil–Water Removal

Recent research has mainly focused on changing the hydrophilic surfaces of PU sponges and MS to hydrophobic surfaces. In the process of separating oil from water, a surface like this gives the sponge excellent oil adsorption and good water repellency.

It is essential to create a quick and effective way of adjusting the hydrophilic sponge surface to the hydrophobic sponge surface in order to satisfy the application requirements of eliminating oils and solvents. The common synthesis techniques mentioned in the literature, such as dip coating, chemical vapor deposition, *in situ* chemical reaction, and carbonization, are summarized in this section. The most popular approach for creating the desired sponge is dip coating.

In the dip-coating process, the sponge is immersed in a solution containing modified materials for several times, followed by a drying process, resulting in a hydrophobic sponge. Organosilicon-modified sponges have typically been prepared by a dip-coating process involving a silanized melamine sponge and octadecyltrichlorosilane. The produced sponges exhibited excellent superhydrophobicity, high adsorption capacity up to 163 g/g for chloroform, excellent recyclability, and good retention of adsorption capacity (Peng et al., 2019).

In recent years, polymerization has become a widely used method for making superhydrophobic sponges. The most popular cross-linking substance is dopamine, which self-polymerizes to produce strong covalent or noncovalent forces with various materials.

Wang et al. prepared PU sponges reinforced with superhydrophobic and superlipophilic CNTs with dopamine oxidative self-polymerization. The resulting sponges exhibited an oil adsorption capacity of up to 34.5 g/g and retained high adsorption performance after being reused 150 times to remove oil from oil–water mixtures. Furfuryl alcohol (FA) is not only miscible with water and many organic solvents but can be polymerized to hydrophobic poly(furfuryl alcohol) (PFA) under a variety of harsh conditions, making it a viable raw material (Yao & Wang, 2007).

### 7.3.2.4 $TiO_2$-based Nanomaterials for Oil–Water Removal

The functionalized material here was successfully produced in a simple, environmentally friendly, and inexpensive manner by decorating the surface of cotton with chemical moieties containing photocatalytic $TiO_2$ nanoparticles and pH-sensitive carboxyl groups. Even more interesting is the use of the photocatalytic properties of $TiO_2$. This produced cotton can also be used to purify water by breaking down water-soluble dyes. Combining the surface nano-roughness structure with appropriate surface ingredients gives the resulting cotton excellent properties of tunable surface wettability, photocatalytic degradation, and durability.

Due to its enhanced delicate wettability and durability, the resulting cotton can be quickly and efficiently vacuum-driven *in situ* to remove oil from wastewater. More importantly, the obtained super-wet cotton shows positive UV degradation performance toward water-soluble organic contaminants. This is facilitated by the photocatalytic properties of $TiO_2$. The developed cotton is seen as a very promising option for the treatment of oily wastewater, whether the mixture is free or emulsified, in a simple, safe, and time-saving way to eliminate organic contaminants. It not only removes but also provides an attractive alternative for oil recovery (Qu et al., 2021).

### 7.3.3 Polymeric Materials Used for Bioremediation

Hydrogenation, ion exchange, liquid–liquid extraction, activated carbon adsorption, forward and inverse osmosis, electrolysis, sonochemistry, UV irradiation, and oxidation are some of the current multistep water purification techniques. Some processes are not always economical since they need expensive catalysts like platinum, palladium, rhodium, and gold nanoparticles on carbon or other supports, high temperatures, and high pressures (e.g., hydrogenation and hydrodechlorination). Some substances can create more dangerous or mutagenic derivatives. For this reason, several bioremediation polymers have recently been introduced. For the effective immobilization of microbes and cells, a variety of synthetic and natural polymers with diverse functional groups have been utilized successfully. Cryogelation is a method for forming macroporous polymeric gels at extremely low temperatures. To create scaffolds or supports for immobilizing bacterial, viral, and other cells, macroporous hydrogels have been employed. Synthetic polymers (polyvinyl alcohol (PVA), polyethylene-glycol (PEG), poly-acrylamide (PAM), polypropylene (PP), polycarbamoyl sulphonate (PCS), polyvinylchloride (PVC), polyethylene (PE), polyacrylonitrile (PAN), and PU) are utilized for the preparation of inexpensive, nontoxic, and potentially reactive functional group carriers for entrapment of microorganisms (Berillo et al., 2021). The employment of membrane technology to remove particles, bacteria, and colloidal materials from effluent streams has recently attracted growing interest. As pores in the ultrafiltration skin-layer fall within the 10–50 nm size range and components causing color and turbidity may be eliminated, ultrafiltration can particularly replace conventional physicochemical methods of clearing and disinfection. The Institute for Polymer Science (Stellenbosch University, South Africa) provided anisotropic internally skinned polysulfone capillary membranes. Wet-phase inversion was used to create membranes in accordance with the Jacobs et al. fabrication methodology. The capillary membrane has the following measurements: 110 mm length, 0.7 mm internal diameter, and 1.2 mm external diameter. The bioreactor's total effective membrane area was 4.838 cm² (Edwards et al., 1999).

### 7.3.3.1 Silica-based Functionalized Polymer

In replacement for conventional wet-chemistry techniques, plasma polymerization is a green process that can be used to synthesize heavy metal adsorbents. With this technique, virtually all solid materials can be altered with low or no surface preparation needed. Since there is no solvent used in the process, almost no waste is generated. In this method, a precursor liquid monomer is initially turned into vapor and then, under the influence of an electric field, stimulated into the plasma. The partially ionized gas plasma, also known as the fourth state of matter, is made up of neutral atoms and molecules, electrons, and charged particles. Any surface exposed to plasma will have the broken monomer species reassemble to form a nanometer-thin layer of plasma polymer film. Plasma polymer films cling to a variety of surfaces and are highly cross-linked. Because of these benefits, plasma polymerization technique can be used to surface-functionalize particles of any size, shape, or surface chemistry. Due to the huge surface area in contact with the plasma, particle surface modification is more difficult than surface modification on planar surfaces. Specific reactor designs, such as revolving, fluidized bed, particle injection, and magnetic stirrer, have been used earlier on to achieve a homogenous distribution of functions onto particles. As a result of the plasma polymerization technology, our prior research demonstrated the successful production of hydrocarbon- and amine-functionalized silica particles for the removal of hydrophobic and negatively charged impurities, respectively (Akhavan et al., 2015).

### 7.3.3.2 Amine-Functionalized Polymers

Chitin, one of the most prevalent biopolymers in nature, can be used to make chitosan. Primary amino groups in chitosan can be further functionalized using a variety of chemical ligands. Chitosan resins were created by Oshita et al. (2009) and modified using a variety of chelating agents, including iminodiacetic acid (IDA). Several transition metals, such as Co(II), Ni(II), Cu(II), Zn(II), Cd(II); lanthanides, such as Ce(III), Nd(III), and Eu(III); and oxo-acids, such as V(V), Ge(IV), and Mo(VI), were effectively removed.

Repo et al. (2013) found that aminopolycarboxylic acids (APCAs) have chelating capabilities that can be used to remove metals from contaminated water. IDA, nitrilotriacetic acid (NTA), ethylenediaminetetraacetic acid (EDTA), and diethylenetriaminepentaacetic acid (DTPA) are the four most popular APCAs that self-prepared adsorbents can be functionalized with polymers. APCAs are a significant class of chelating substances with the capacity to bind multiple carboxylate groups to one or more nitrogen atoms. They are also used to stop metal precipitates from forming. The surface functionalization of these high affinity binding groups has attracted a lot of scientific attention recently. The most widely utilized APCAs for binding metal ions are NTA, IDA, DTPA, and EDTA. The usage of APCAs is widespread in various industrial processes and products, such as in (i) the prevention of metal precipitation, (ii) the inhibition of metal ion catalysis, (iii) the removal of metal ions, or (iv) the maintenance of metal ions in the solution. Detergents, prescription medications, cosmetics, and food items all include APCA ingredients. The primary usage of APCAs in industrial processes is as cleaning agents. The simplest APCA, IDA has two carboxyl groups. In recent decades, IDA-functionalized adsorbents have been the subject of extensive research. The amorphous form of silicon dioxide, known as silica gel, is composed of silicon atoms connected by siloxane bonds to oxygen atoms. The hydroxyl groups on the surface of silica gel can be further functionalized utilizing the well-known silanization process. After reacting the iodo group with diethyliminodiacetate and hydrolyzing it, 3-iodopropyltrimethoxysilane was used as the silylating agent to create IDA-functionalized silica gel. This substance has been used to successfully remove metals from synthetic wastewater, including Co(II), Ni(II), Cu(II), and Zn(II).

### 7.3.4 Membrane Technology in Heavy Metal Removal

A lot of research has been done on membrane filtration technology in particular because it can offer high removal efficiency, easy fabrication and operation, space savings, and no phase change.

The majority of nanofiltration (NF) membranes used today are thin-film composites (TFCs) made of an asymmetrically porous substrate and a very thin polymeric film. The ultrathin polymeric film achieves the rejection of diverse dissolved species, and the porous substrate of the TFC membrane provides the mechanical strength required for the membrane to endure the high operating pressure. The molecularly created Nexar copolymer, which is composed of pentablock copolymer poly(tert-butylstyrene-b-hydrogenated isoprene-b-sulfonated styrene-b-tert-butylstyrene) (tBS-HI-SS-HI-tBS), has a number of beneficial properties. First, it may sulfonate to a high degree without significantly losing mechanical stability. Second, the copolymer's level of sulfonation can be easily adjusted for different uses. Third, the central sulfonated styrene block merging to develop a continuous water transport channel throughout the film creation process contributes to further improve the membrane's water permeability (Thong et al., 2014).

### 7.3.4.1 Polyacrylonitrile Membrane

Polyacrylonitrile (PAN) is a good candidate for fabricating the porous substrate because it has strong acid resistance against a variety of inorganic acids, including hydrochloric acid, nitric acid, and sulfuric acid, as well as because it is less expensive than other polymers like polysulfone and polyimide. This is necessary to meet the strong acid condition during the heavy metal removal process. But inevitably, the PAN substrate-based TFC membrane frequently has poor adhesion between the substrate and the selective layer. The resulting membrane might be highly sensitive to swelling in an acidic environment, which would severely reduce rejection. Many attempts have been made to develop a cushion layer to enhance the interaction between the substrate and the interfacial polymerized selective layer, such as the extensively researched polydopamine coating layer. However, adding more manufacturing stages may result in increased cost and repeatability problems when the composite membranes are scaled up (Jia et al., 2019).

### 7.3.4.2 Sulfonated Pentablock Copolymer Membrane Addition of Graphene Oxide

The removal of heavy metals from water was explored using sulfonated pentablock copolymer (s-PBC, Nexar) membranes and s-PBC/graphene oxide (GO) nanocomposite membranes. Drop casting was used to create the membranes, and their chemical, structural, and morphological characteristics were examined using SEM, FT-IR spectroscopy, dynamic mechanical analysis (DMA), and small-angle X-ray scattering (SAXS), among other techniques. The polymer and the s-PBC/GO nanocomposite's adsorption capacities and kinetics were examined for the removal of several heavy metal ions ($Ni^{2+}$, $Co^{2+}$, $Cr^{3+}$, and $Pb^{2+}$) from aqueous solutions containing the corresponding metal salts at various concentrations. The examined s-PBC membrane exhibits good efficiency as a result of the presence of sulfonic groups, which are crucial to the metal ion adsorption process (Filice et al., 2020).

### 7.3.4.3 Plasma-Modified Membranes Used for Oil–Water Separation

Since it will not damage the surface polymeric materials, even temperature-sensitive materials, nonthermal plasma is typically utilized to change the surface of materials. A few kilohertz to several megahertz can be maintained as the frequency of the applied voltage. Increasing the superhydrophilicity or superhydrophobicity of nanomaterials was frequently done in order to develop extremely effective NF membranes (Manakhov et al., 2022).

A prospective technique called NF offers a combined approach to rejecting both organic and inorganic contaminants. Due to its ability to produce clean water from greasy solutions, NF has a significant potential for oil–water separation. If the produced water discharge is selected, NF permeate may also be sufficiently clean to satisfy the strict discharge standards. The widespread use of NF is hindered by a number of issues, including economic feasibility, robustness, and moderate efficiency. The most cutting-edge environmentally friendly oil–water separation techniques based on modified NF membranes are outlined in this section. Although excellent separation performance for water-in-oil emulsions stabilized by surfactants with up to 99% filtration efficiency and > 99.98% oil recovery purity has been found, it is not economically viable for widespread use (Park & Barnett, 2001).

### References

Akhavan, B., Jarvis, K., & Majewski, P. (2015). Plasma polymer-functionalized silica particles for heavy metals removal. *ACS Applied Materials & Interfaces*, 7(7), 4265–4274. https://doi.org/10.1021/am508637k

Ali, A., Haq, I. U., Akhtar, J., Sher, M., Ahmed, N., & Zia, M. (2017). Synthesis of Ag-NPs impregnated cellulose composite material: Its possible role in wound healing and photocatalysis. *IET Nanobiotechnology*, 11(4), 477–484. https://doi.org/10.1049/iet-nbt.2016.0086

Al-Qahtani, K. M. (2017). Cadmium removal from aqueous solution by green synthesis zero valent silver nanoparticles with *Benjamina* leaves extract. *The Egyptian Journal of Aquatic Research*, 43(4), 269–274. https://doi.org/10.1016/j.ejar.2017.10.003

Berillo, D., Al-Jwaid, A., & Caplin, J. (2021). Polymeric materials used for immobilisation of bacteria for the bioremediation of contaminants in water. *Polymers*, 13(7), Article 7. https://doi.org/10.3390/polym13071073

Bhanjana, G., Dilbaghi, N., Kim, K.-H., & Kumar, S. (2017). Carbon nanotubes as sorbent material for removal of cadmium. *Journal of Molecular Liquids*, 242, 966–970. https://doi.org/10.1016/j.molliq.2017.07.072

Chauhan, G., González-González, R. B., & Iqbal, H. M. N. (2022). Bioremediation and decontamination potentials of metallic nanoparticles loaded nanohybrid matrices—A review. *Environmental Research*, 204, 112407. https://doi.org/10.1016/j.envres.2021.112407

Cumbal, L., & SenGupta, A. K. (2005). Arsenic removal using polymer-supported hydrated iron(III) oxide nanoparticles: Role of Donnan membrane effect. *Environmental Science & Technology*, 39(17), 6508–6515. https://doi.org/10.1021/es050175e

Das, S. K., Khan, M. M. R., Parandhaman, T., Laffir, F., Guha, A. K., Sekaran, G., & Baran Mandal, A. (2013). Nano-silica fabricated with silver nanoparticles: Antifouling adsorbent for efficient dye removal, effective water disinfection and biofouling control. *Nanoscale, 5*(12), 5549–5560. https://doi.org/10.1039/C3NR00856H

Djellabi, R., Yang, B., Adeel Sharif, H. M., Zhang, J., Ali, J., & Zhao, X. (2019). Sustainable and easy recoverable magnetic $TiO_2$-Lignocellulosic Biomass@$Fe_3O_4$ for solar photocatalytic water remediation. *Journal of Cleaner Production, 233*, 841–847. https://doi.org/10.1016/j.jclepro.2019.06.125

Dunphy Guzman, K. A., Finnegan, M. P., & Banfield, J. F. (2006). Influence of surface potential on aggregation and transport of titania nanoparticles. *Environmental Science & Technology, 40*(24), 7688–7693. https://doi.org/10.1021/es060847g

Edwards, W., Leukes, W. D., Rose, P. D., & Burton, S. G. (1999). Immobilization of polyphenol oxidase on chitosan-coated polysulphone capillary membranes for improved phenolic effluent bioremediation. *Enzyme and Microbial Technology, 25*(8), 769–773. https://doi.org/10.1016/S0141-0229(99)00116-7

Feng, L., Cao, M., Ma, X., Zhu, Y., & Hu, C. (2012). Superparamagnetic high-surface-area Fe3O4 nanoparticles as adsorbents for arsenic removal. *Journal of Hazardous Materials, 217*, 439–446. https://doi.org/10.1016/j.jhazmat.2012.03.073

Filice, S., Mazurkiewicz-Pawlicka, M., Malolepszy, A., Stobinski, L., Kwiatkowski, R., Boczkowska, A., Gradon, L., & Scalese, S. (2020). Sulfonated pentablock copolymer membranes and graphene oxide addition for efficient removal of metal ions from water. *Nanomaterials, 10*(6), Article 6. https://doi.org/10.3390/nano10061157

Hawkes, S. J. (1997). What is a "heavy metal"? *Journal of Chemical Education, 74*(11), 1374. https://doi.org/10.1021/ed074p1374

Jia, T.-Z., Lu, J.-P., Cheng, X.-Y., Xia, Q.-C., Cao, X.-L., Wang, Y., Xing, W., & Sun, S.-P. (2019). Surface enriched sulfonated polyarylene ether benzonitrile (SPEB) that enhances heavy metal removal from polyacrylonitrile (PAN) thin-film composite nanofiltration membranes. *Journal of Membrane Science, 580*, 214–223. https://doi.org/10.1016/j.memsci.2019.03.015

Kaur, S., & Roy, A. (2021). Bioremediation of heavy metals from wastewater using nanomaterials. *Environment, Development and Sustainability, 23*(7), 9617–9640. https://doi.org/10.1007/s10668-020-01078-1

Kulshreshtha, A., Agrawal, R., Barar, M., & Saxena, S. (2014). A review on bioremediation of heavy metals in contaminated water. *IOSR Journal of Environmental Science, Toxicology and Food Technology, 8*, 44–50. https://doi.org/10.9790/2402-08714450

Malik, S., Kishore, S., Shah, M. P., & Kumar, S. A. (2022). A comprehensive review on nanobiotechnology for bioremediation of heavy metals from wastewater. *Journal of Basic Microbiology, 62*(3–4), 361–375. https://doi.org/10.1002/jobm.202100555

Manakhov, A., Orlov, M., Grokhovsky, V., AlGhunaimi, F. I., & Ayirala, S. (2022). Functionalized nanomembranes and plasma technologies for produced water treatment: A Review. *Polymers, 14*(9), Article 9. https://doi.org/10.3390/polym14091785

Oshita, K., Sabarudin, A., Takayanagi, T., Oshima, M., & Motomizu, S. (2009). Adsorption behavior of uranium(VI) and other ionic species on cross-linked chitosan resins modified with chelating moieties. *Talanta, 79*(4), 1031–1035. https://doi.org/10.1016/j.talanta.2009.03.035

Park, E., & Barnett, S. M. (2001). Oil/water separation using nanofiltration membrane technology. *Separation Science and Technology, 36*(7), 1527–1542. https://doi.org/10.1081/SS-100103886

Peng, M., Zhu, Y., Li, H., He, K., Zeng, G., Chen, A., Huang, Z., Huang, T., Yuan, L., & Chen, G. (2019). Synthesis and application of modified commercial sponges for oil-water separation. *Chemical Engineering Journal, 373*, 213–226. https://doi.org/10.1016/j.cej.2019.05.013

Pete, A. J., Bharti, B., & Benton, M. G. (2021). Nano-enhanced bioremediation for oil spills: A review. *ACS ES&T Engineering, 1*(6), 928–946. https://doi.org/10.1021/acsestengg.0c00217

Qu, M., Liu, Q., Yuan, S., Yang, X., Yang, C., Li, J., Liu, L., Peng, L., & He, J. (2021). Facile fabrication of $TiO_2$-functionalized material with tunable superwettability for continuous and controllable oil/water separation, emulsified oil purification, and hazardous organics photodegradation. *Colloids and Surfaces A: Physicochemical and Engineering Aspects, 610*, 125942. https://doi.org/10.1016/j.colsurfa.2020.125942

Rando, G., Sfameni, S., Galletta, M., Drommi, D., Cappello, S., & Plutino, M. R. (2022). Functional nanohybrids and nanocomposites development for the removal of environmental pollutants and bioremediation. *Molecules, 27*(15), Article 15. https://doi.org/10.3390/molecules27154856

Rangel-Muñoz, N., González-Barrios, A. F., Pradilla, D., Osma, J. F., & Cruz, J. C. (2020). Novel bionanocompounds: Outer membrane protein A and laccase co-immobilized on magnetite nanoparticles for produced water treatment. *Nanomaterials, 10*(11), Article 11. https://doi.org/10.3390/nano10112278

Recillas, S., García, A., González, E., Casals, E., Puntes, V., Sánchez, A., & Font, X. (2011). Use of $CeO_2$, $TiO_2$ and $Fe_3O_4$ nanoparticles for the removal of lead from water: Toxicity of nanoparticles and derived compounds. *Desalination, 277*(1), 213–220. https://doi.org/10.1016/j.desal.2011.04.036

Repo, E., Warchoł, J. K., Bhatnagar, A., Mudhoo, A., & Sillanpää, M. (2013). Aminopolycarboxylic acid functionalized adsorbents for heavy metals removal from water. *Water Research, 47*(14), 4812–4832. https://doi.org/10.1016/j.watres.2013.06.020

Sharma, U., & Sharma, J. G. (2022). Nanotechnology for the bioremediation of heavy metals and metalloids. *Journal of Applied Biology & Biotechnology*, 34–44. https://doi.org/10.7324/JABB.2022.100504

Shi, Q., Vitchuli, N., Nowak, J., Noar, J., Caldwell, J. M., Breidt, F., Bourham, M., McCord, M., & Zhang, X. (2011). One-step synthesis of silver nanoparticle-filled nylon 6 nanofibers and their antibacterial properties. *Journal of Materials Chemistry, 21*(28), 10330. https://doi.org/10.1039/c1jm11492a

Simonsen, G., Strand, M., & Øye, G. (2018). Potential applications of magnetic nanoparticles within separation in the petroleum industry. *Journal of Petroleum Science and Engineering, 165*, 488–495. https://doi.org/10.1016/j.petrol.2018.02.048

Sudhakar, M. S., Aggarwal, A., & Sah, M. K. (2020). Chapter 14—Engineering biomaterials for the bioremediation: Advances in nanotechnological approaches for heavy metals removal from natural resources. In M. P. Shah, S. Rodriguez-Couto, & S. S. Şengör (Eds.), *Emerging technologies in environmental bioremediation* (pp. 323–339). Elsevier. https://doi.org/10.1016/B978-0-12-819860-5.00014-6

Thong, Z., Han, G., Cui, Y., Gao, J., Chung, T.-S., Chan, S. Y., & Wei, S. (2014). Novel nanofiltration membranes consisting of a sulfonated pentablock copolymer rejection layer for heavy metal removal. *Environmental Science & Technology, 48*(23), 13880–13887. https://doi.org/10.1021/es5031239

Wali, L. A., Alwan, A. M., Dheyab, A. B., & Hashim, D. A. (2019). Excellent fabrication of Pd-Ag NPs/PSi photocatalyst based on bimetallic nanoparticles for improving methylene blue photocatalytic degradation. *Optik, 179*, 708–717. https://doi.org/10.1016/j.ijleo.2018.11.011

Wang, H., Zhao, Q., Zhang, K., Wang, F., Zhi, J., & Shan, C.-X. (2022). Superhydrophobic nanodiamond-functionalized melamine sponge for oil/water separation. *Langmuir*, *38*(37), 11304–11313. https://doi.org/10.1021/acs.langmuir.2c01480

Wang, J., Zheng, S., Shao, Y., Liu, J., Xu, Z., & Zhu, D. (2010). Amino-functionalized $Fe_3O_4$@$SiO_2$ core–shell magnetic nanomaterial as a novel adsorbent for aqueous heavy metals removal. *Journal of Colloid and Interface Science*, *349*(1), 293–299. https://doi.org/10.1016/j.jcis.2010.05.010

Xin, X., Wei, Q., Yang, J., Yan, L., Feng, R., Chen, G., Du, B., & Li, H. (2012). Highly efficient removal of heavy metal ions by amine-functionalized mesoporous $Fe_3O_4$ nanoparticles. *Chemical Engineering Journal*, *184*, 132–140. https://doi.org/10.1016/j.cej.2012.01.016

Yao, J., & Wang, H. (2007). Preparation of crystalline mesoporous titania using furfuryl alcohol as polymerizable solvent. *Industrial & Engineering Chemistry Research*, *46*(19), 6264–6268. https://doi.org/10.1021/ie070319t

Zhang, S., Niu, H., Cai, Y., Zhao, X., & Shi, Y. (2010). Arsenite and arsenate adsorption on coprecipitated bimetal oxide magnetic nanomaterials: $MnFe_2O_4$ and $CoFe_2O_4$. *Chemical Engineering Journal*, *158*(3), 599–607. https://doi.org/10.1016/j.cej.2010.02.013

Zhang, T., Gu, B., Qiu, F., Peng, X., Yue, X., & Yang, D. (2018). Preparation of carbon nanotubes/polyurethane hybrids as a synergistic absorbent for efficient oil/water separation. *Fibers and Polymers*, *19*(10), 2195–2202. https://doi.org/10.1007/s12221-018-8399-1

# 8

## Energy Applications of Functional Nanomaterials and Polymers

### 8.1 Brief Overview of Functional Nanomaterials and Polymers in Energy Applications

Significant interest has been drawn toward green energy amid the global energy crisis caused by utilizing nonrenewable resources. The current scenario of energy relying on nonrenewable services has significantly deteriorated our motherly earth and is also approaching its exhaustion. Hence, deriving energy from renewable resources can help overcome these issues. The future of energy is heavily dependent on environment-friendly resources and technologies.

Numerous kinds of research are conducted in the field of the energy sector to develop or improvise the existing technologies in pursuit of an efficient and sustainable system. Some technologies use nanomaterials and polymers to enhance their efficiency and cost reduction. Nanomaterials and polymers play a considerable role in energy applications. They can be used as a catalyst in fuel cells, energy conversion, membrane fabrication, energy storage, and many more. The exciting features of nanomaterials and polymers are high surface area, improved conductivity, tailored size and shape, enhanced durability, and tunable surface chemistry. To broaden the range of applications, the functionalization of nanomaterials and polymers has been introduced. It is functionalized with other particles to improve its efficacy and introduce new functionalities (Figure 8.1). Several functional groups can be added to nanomaterials and polymers, and below mentioned are a few examples. Carboxylic acid (-COOH) and phosphonic acid ($-PO_3H_2$) assist nanomaterials to disperse and stabilize in solvents and water. Moreover, it can make it easier for nanomaterials to adhere to other molecules. The amine ($-NH_2$) functional group can make nanomaterials and polymers more catalytically active. Moreover, it can increase a material's electrical conductivity and speed up the process. The functional group sulfonic acid ($-SO_3H$) can enhance the hydrophilicity and surface charge of nanomaterials. Moreover, it can make it easier for nanomaterials to bond to metal ions and surfaces. Ester (-COO) helps nanoparticles dissolve and remain stable in solvents (Figure 8.2). Moreover, it can help cross-linked polymer networks form (Makvandi et al., 2021).

DOI: 10.12019781003391364-8

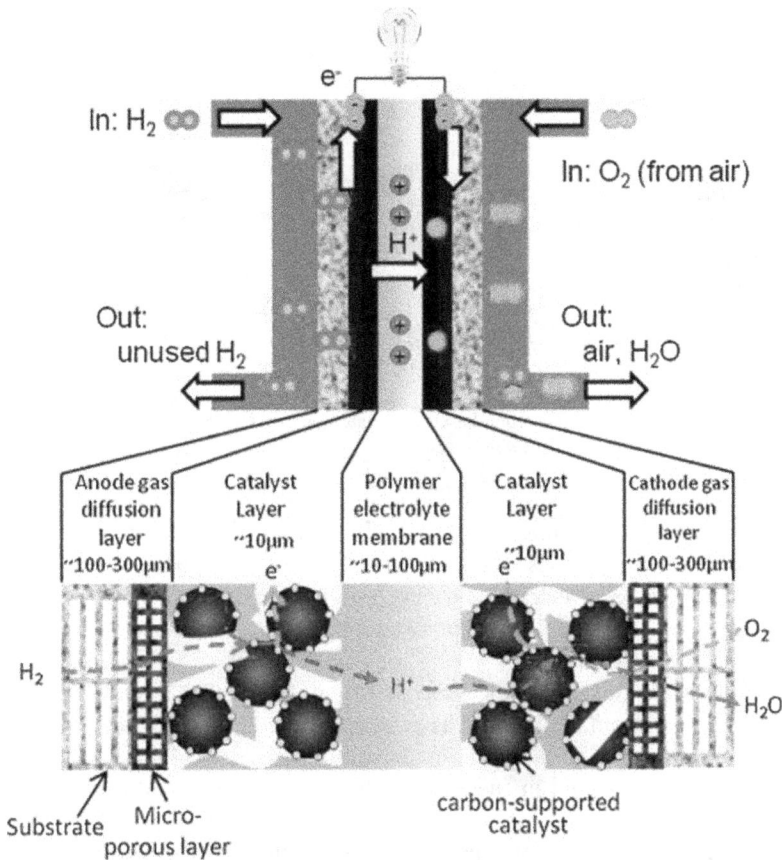

**FIGURE 8.1**
Schematical representation of a polyelectrolyte fuel cell setup (Mishler et al., 2012). *Reproduced with permission from Elsevier.*

Based on the required properties and uses and the compatibility of that functional group to the neat materials, we can select the required functional group for the functionalization. The covalent, noncovalent, sol-gel method, electrochemical, and plasma functionalization are some of the functionalization processes.

## 8.2 Application of Functionalized Nanomaterials and Polymers in Energy

### 8.2.1 Fuel Cells

Because of their unique characteristics, fuel cells have been the focus of new energy technology research. They are electrochemical devices that use an electrocatalytic

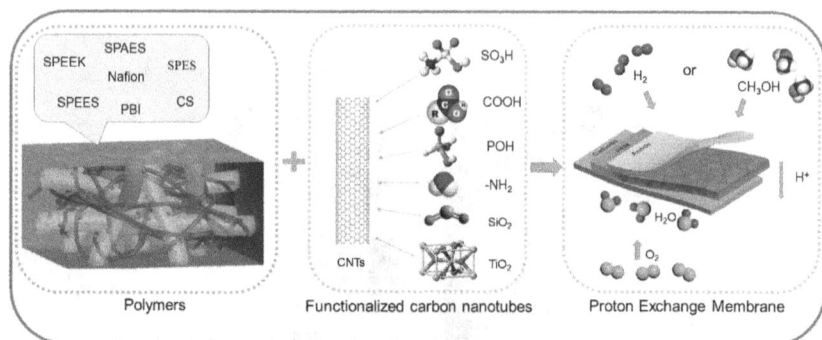

**FIGURE 8.2**

Functionalization of polymers and carbon-based nanomaterials for a membrane-based fuel cell (Gao et al., 2023). *Reproduced with permission from Elsevier.*

process to transfer the chemical energy stored in fuels directly into electrical currents. In contrast to batteries, fuel cells are flexible, have a high theoretical efficiency, do not require recharging, and can constantly produce power as long as they are fed with fuel. Functionalized nanomaterials and polymers have many valuable applications in fuel cells.

Some inorganic nanomaterials used in fuel cells are $TiO_2$, $Al_2O_3$, $SiO_2$, and $Bi_2WO_6$. Carbon nanomaterials such as graphene, fullerenes, and CNTs are utilized for membrane fabrication. Nanocellulose, metal-organic, and carbon organic frameworks have also been used in fuel cells. Polymers such as PES, PEEK, PSF, and PPSU, are used in fuel cells.

Catalysts: Fuel cells can use nanomaterials (such as platinum nanoparticles, which are the most commonly used) as catalysts to speed up the electrochemical reactions. Although they have high catalytic activity, the expensive cost and challenges in commercialization hindered their application in fuel cells. Graphene-based materials, such as graphene oxide (GO), functionalized with nitrogen-containing molecules can work with platinum as co-catalysts to improve fuel cell performance while using less platinum (Su & Hu, 2021). The utilization of carbon nanotubes (CNTs) functionalized with nitrogen-containing molecules has exhibited increased catalytic activity and stability compared with conventional platinum-based catalysts (Wong et al., 2012).

Liew et al. (2015) researched manganese oxide/functionalized CNT nanocomposites as catalyst for oxygen reduction reaction (ORR) in a microbial fuel cell. The electron transfer process was improved, and the ORR was made possible by the unique interaction between $MnO_2$ and f-CNTs (functionalized carbon nanotubes). The *in situ* hydrothermal synthesis approach was used to create the $MnO_2$/f-CNT nanocomposite. The results of the test on microbial fuel cells revealed that $MnO_2$/f-CNTs had a greater power density (520 mW m²) than CNTs (275 mW m²) and f-CNTs (440 mW m²). Moreover, the microbial fuel cell (MFC) with $MnO_2$/f-CNT catalyst had the highest coulombic efficiency (28.65%) and COD removal value (86.6%) of the three investigated catalysts. This study concluded that $MnO_2/f$- CNTs can be a helpful catalyst in applying microbial fuel cells. CNTs have been widely used to support Pt catalysts in fuel cell applications due to their extraordinary electrical, mechanical, and structural properties. The raw CNTs were functionalized using a conventional sonochemical treatment method. The synthesis of Pt on FCNTs and RCNTs was prepared by a microwave-assisted technique. Nafion membrane was used in the cell. The Pt catalysts incorporated in f-CNTs had shown significantly enhanced electrochemically active catalytic surface area and PEM fuel cell performance compared to Pt/RCNTs and Pt/C. The sonochemical method removed impurities on CNTs and allowed the tubes to be functionalized, giving a pathway to host Pt metal ions (W. Zhang et al., 2010). To enhance the efficiency of the fuel cell, nanomaterials can be created to have particular qualities, such as more significant surface area or higher catalytic activity.

Electrolytes: In fuel cells, polymers can be used as solid or polymer electrolytes. Solid polymer electrolytes can provide superior ion conductivity and durability compared to liquid electrolytes. Nanocomposite polymer electrolytes can also be created by enhancing the polymer matrix with valuable nanoparticles. Polymer membranes are an essential component of fuel cells, separating the anode and cathode compartments while allowing for the transport of protons or other charged particles. Functional nanomaterials can be incorporated into polymer membranes to improve their proton conductivity and durability. The most commonly used membrane is Nafion, which has high ionic conductivity, better selectivity, and low gas permeability. But it was found that it had low functionality at elevated temperatures.

Gouda et al. (2021) fabricated nanocomposite membranes for direct borohydride fuel cells, low-temperature fuel cells from a ternary polymer blend of poly(vinyl alcohol), poly(vinyl pyrrolidone), and poly(ethylene oxide) incorporated with $SO_4$-$TiO_2$ nanotubes and $PO_4$-$TiO_2$ as doping agents. The impregnation–calcination method was used for the functionalization of $TiO_2$. It was found that the doped membranes reduced the crossover of borohydride ($BH_4$-), and permeability decreased with the increasing content of doping agents. With the performance of the fuel cell, the doped membranes showed slightly lower open-circuit voltage compared to Nafion. The power density was almost near to the value generated by Nafion. So, these doped membranes showed the potential to replace Nafion in the application of direct borohydride fuel cells (DBFCs) (Figure 8.3).

Branchi et al. (2015) functionalized $Al_2O_3$ and dispersed it in Nafion, which was used as an electrolyte–proton exchange membrane fuel cell (PEMFC). This composite membrane improved the performance of the fuel cell compared to the neat Nafion membrane at high temperatures. But these composite membranes showed lower ionic exchange capacity and water uptake than the Nafion membrane.

Gas diffusion layers (GDLs): Interdependent characteristics such as water management, porosity, and the graded structure of the GDL have a significant impact on the power performance of the PEMFC. The GDL should have balanced and integrated hydrophobic (water expelling) and hydrophilic qualities (water retaining). Careful balancing of these factors is necessary to ensure that the fuel cell system

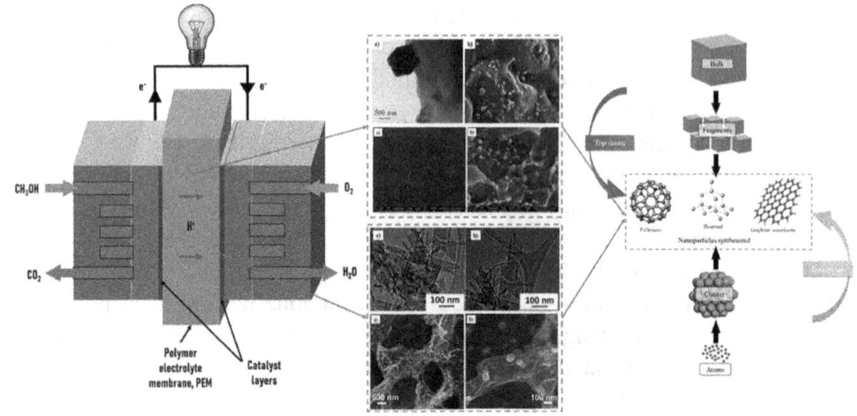

**FIGURE 8.3**

Schematical representation of a fuel cell work station and enhancing of cell performance with a functionalized polymeric membrane (Raduwan et al., 2022). *Reproduced with permission from Elsevier.*

operates without flooding and high humidity. Carbon materials such as graphene and CNTs have been used to fabricate microporous layers (Jha et al., 2013).

## 8.2.2 Solar Cells

A solar cell, also known as a photovoltaic cell, is a technological innovation that uses the physical and chemical phenomenon known as the photovoltaic effect to convert light energy directly into electricity. When government services cannot be provided, or solar energy becomes more affordable than other forms of energy, solar cells become crucial in creating clean and sustainable electric power. Although silicon solar cells are popular, extensive research is being done to develop less expensive solar cells, such as polymer solar cells and perovskite solar cells. Nanomaterials and polymers have been used in solar cell components, which play a pivotal role in energy production.

A wider variety of light wavelengths can be absorbed by functionalized materials, which increases their ability to convert solar energy into electricity. It is feasible to increase the capacity of materials to transport electrons, which could result in higher solar cell efficiency. Functionalized materials are sometimes more affordable than standard materials, making them a more economical choice for manufacturing solar cells. The next-generation solar cells may benefit from using functionalized nanomaterials and polymers due to their potential for greater efficiency, increased stability, and lower costs.

In solar cells, nanoparticles like quantum dots can be used as an absorbing photovoltaic material. Bulky materials such as silicon, copper indium gallium selenide, and cadmium telluride could be replaced by quantum dots. These nanoparticles can be made more effective by surface ligand functionalization, which improves the stability and electrical characteristics of the particles. A photosensitive dye is utilized in dye-sensitized solar cells (DSSCs) to absorb sunlight and produce energy. It is possible to functionalize nanoparticles like $TiO_2$ with dyes that have a high absorption coefficient, improving light harvesting and effectiveness (Figure 8.4). In this method, the pigment molecule can be replaced by a polymer to produce charge carriers and absorb light (Kishore Kumar et al., 2020). A promising technology that has demonstrated great efficiency is perovskite solar cells. Perovskite can be functionalized with nanomaterials like $TiO_2$ and ZnO to enhance the interface between the perovskite layer and the electron transport layer, which will improve charge transfer and increase efficiency. This method enhances the perovskite solar cell's efficiency by using a polymer as a layer that transports holes. Gold and silver plasmonic nanoparticles can improve the solar cells' capacity to absorb light. The performance of the nanoparticles can be enhanced by functionalizing them with ligands that increase their stability and dispersibility in the solar cell. Light-absorbing nanowires made of silicon and zinc oxide can be used in solar cells (Hussain et al., 2018). Bulk heterojunction solar cells (BHJ-SCs) have a thin-film active layer by combining the polymer with a fullerene variant or a non-fullerene acceptor. Light captured by the polymer produces an exciting, which the acceptor can then divide into free charges (holes and electrons). The electrodes then collect

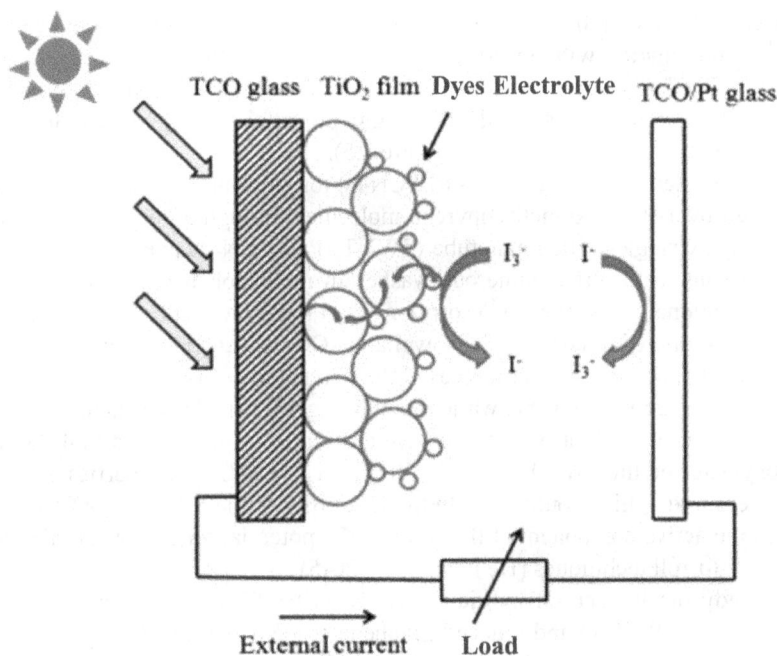

**FIGURE 8.4**
Schematic illustration of a dye-sensitized solar cell (DSSC) (Gong et al., 2012). *Reproduced with permission from Elsevier.*

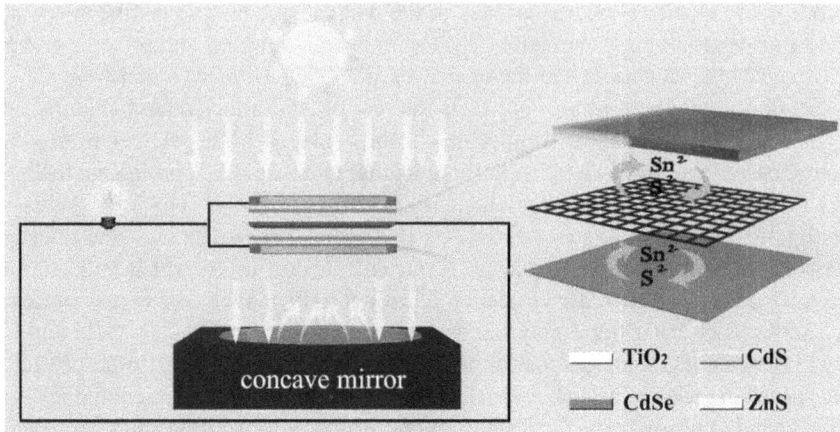

**FIGURE 8.5**
Quantum dot-sensitized solar cells (Xu et al., 2020). *Reproduced with permission from Springer.*

the free charges, producing an electrical current (Nelson, 2011). In this method, a polymer solar cell can be used in addition to an inorganic solar cell to increase the device's efficiency. To profit from both, hybrid solar cells combine organic and inorganic components. In this method, a polymer can be used as a p-type semiconductor layer in combination with an n-type inorganic material to create a heterojunction solar cell (Niederhausen et al., 2021). DSSCs, BHJ-SCs, and quantum dot-sensitized solar cells (QDSSCs) are several emerging photovoltaic cells that have drawn significant attention from researchers (Figure 8.5).

Single-walled carbon nanotubes (SWCNTs) for efficient DSSCs have been functionalized by 1-(hydroxymethyl)pyrene molecules through a noncovalent process. The complex single-walled nanotube (SWNT)/Py-OH/$N_3$ was used in the assembly of a solar cell. After numerous washes in methanol, it was found that they were exceptionally resistant to desorption when these molecules were anchored to SWNTs. Compared to commercially available -COOH functionalized SWNTs, this noncovalent functionalization increased the percentage of reactive sites (hydroxyl group percentage was 6 wt.%) while maintaining SWNT physicochemical characteristics. Also, it made it possible for dye molecules to be attached to 48% of the hydroxyl sites on the SWNT surfaces. The superior electrical properties of the cell were measured, which confirms both the feasibility of the SWNT/Py-OH/$N_3$ complex as the active component of the cell and the potential for industrial fabrication using roll-to-roll techniques (De Filpo et al., 2015).

To study the impacts of hybrid materials on P3HT: PCBM organic solar cell performance, SWCNTs and reduced graphene oxide (rGO) functionalized by zinc phthalocyanine (ZnPc) covalently and noncovalently were introduced in the P3HT: PCBM blend. These hybrid materials have significantly impacted the electrical characteristics of the investigated blends. Electrical conductivity has been seen to increase by around two orders of magnitude (Kadem et al., 2016).

## 8.2.3 Batteries

Voltage, capacity, and energy efficiency are all factors considered when rating batteries, and they all relate to the electrical potential difference between the positive and negative terminals of a battery. Energy density measures how much energy a battery can store in relation to its volume or weight. Batteries are machinery that uses a chemical process to transform chemical energy into electrical energy. From small electronics like cell phones and flashlights to more powerful ones like electric vehicles and renewable energy systems, they power a broad range of devices.

### 8.2.3.1 Uses of Functionalized Nanomaterials in Batteries

By increasing the energy density, power density, and cycle life of batteries, functionalized nanomaterials have the potential to improve their performance significantly. The creation of high-capacity electrode materials is one of the most exciting uses of functionalized nanoparticles in batteries. For instance, increasing the conductive cathode additive with CNTs or GO can significantly increase a battery's capacity and durability. Using nanoscale metal oxides as anode materials, which can enhance a battery's specific capacity and cycling performance, is another potential application. For instance, increasing the energy density of a battery by using tin oxide nanoparticles as the anode material is possible because tin oxide has theoretically greater capacity than traditional graphite anodes. The efficacy of the electrolyte in batteries can also be enhanced by using functionalized nanomaterials. For instance, the stability and safety of a battery can be improved by using ceramic nanoparticles as the electrolyte substance. Improvements in proton conductivity, gas or methanol permeability, and other hybrid membrane characteristics can all be significantly enhanced by incorporating nanoparticles. The efficacy, effectiveness, and safety of batteries could all be greatly enhanced by using functionalized nanomaterials. Research is still being done to improve the performance of nanomaterial-based batteries by creating novel materials and technologies.

### 8.2.3.2 How We Can Use Functionalized Nanomaterials in Batteries?

Batteries can benefit from using functionalized nanomaterials by performing better in terms of energy density, power density, cyclic life, and safety. Here are some ways to use batteries with functionalized nanomaterials. To expand the electrodes' surface area and improve their electrochemical performance, functionalized nanomaterials can be used as electrode materials. For instance, CNTs can be used as a conductive additive to increase lithium-ion batteries' rate capacity and cycle life. Functionalized nanomaterials can enhance the electrolyte's ion transport characteristics. GO can be added to liquids to improve their ionic conductivity. To increase the safety of batteries, functionalized nanoparticles can be used as separator materials. For instance, to improve the separator's thermal resilience and stop thermal runaway in a battery, nanoscale ceramic particles can be added. To increase the conductivity of the conductors, current collectors made of functionalized nanomaterials can be used. CNTs can be used as a current collector to increase the rate capacity and cycle life of lithium-ion batteries. Anode and cathode coatings made of functionalized nanomaterials can boost a battery's electrochemical efficiency. For instance, carbon can be coated onto silicon nanoparticles to increase the stability and volume of the anode in lithium-ion batteries (Halankar et al., 2021).

### 8.2.3.3 Functionalized Polymers in Batteries

By acting as essential elements of the electrolyte or electrode materials, functionalized polymers can be essential in creating batteries. These polymers can have particular characteristics, like high ionic conductivity or stability in harsh environments, which can enhance the functionality, security, and life of batteries. The creation of solid-state buffers is one battery application for functionalized polymers. Solid-state electrolytes can replace traditional liquid electrolytes, which can be dangerous because they are flammable. Solid-state electrolytes with high ionic conductivity, excellent mechanical characteristics, and great stability can be produced using functionalized polymers. Polyethylene oxide, polyacrylonitrile (PAN), and polyvinylidene fluoride are some functionalized polymers that can be applied to this situation. Electrode materials can also benefit from the use of functionalized polymers by performing better. Functionalized polymers, for instance, can be added to electrodes to improve their electrical conductivity or used as binders to improve the binding of active materials to current collectors. The mixture of polyvinylidene fluoride and hexafluoropropylene, carboxymethyl cellulose, and sodium alginate are a few examples of functionalized polymers that can be used for this purpose.

### 8.2.3.4 Use of Functionalized Polymers in Batteries

There are several methods to use functionalized polymers in batteries. Functionalized polymers can be used as electrolyte layers in batteries. Ions can move between the electrode and cathode with the help of electrolyte membranes. The conductivity and selectivity of the membrane can be increased by using functionalized polymers, which will enhance battery efficiency. Additionally, functionalized polymers can serve as cathode and anode components in batteries. The polymer's electrochemical properties can be enhanced by changing its structure, leading to excellent charge storage and discharge rates. Functionalized polymers are suitable for binders in battery cells. Binders enhance the mechanical integrity of the electrodes by holding the active components of the electrodes together. Better binding between the active components and the current collector can be achieved by using functionalized polymers as binders, which will enhance battery performance in general. Additionally, functionalized polymers can be used as coatings for battery separators, which can enhance the separator's wettability and durability and boost battery performance.

Compared to the $TiO_2$-blended nanohybrid polymer electrolyte (NHPE) and the poly(ethylene glycol) methyl ether methacrylate polymer electrolyte, $TiO_2$-grafted NHPE exhibits greater ionic conductivity, a higher lithium-ion transference number, and greater thermal and electrochemical stability. The hybrid nano-structured electrolyte demonstrates outstanding interface stability with the lithium metal anode and promotes uniform $Li^+$ deposition (Ma et al., 2016). According to electrochemical studies, the strong interaction between oxygen functional groups on oxidized bacterial cellulose (BC) and $SiO_2$ with $Li^+$ and polysulfides allows the composite $SiO_2$ on oxidized bacterial cellulose (o-BC/$SiO_2$) separator to produce a smooth $Li^+$ flux, control the $Li^+$ deposition, and stop the polysulfide shuttling process. In comparison to traditional separators, this composite film provided substantially greater stability, lower polarization voltage, and higher Coulombic efficiency when utilized as the separator in the plating and stripping of lithium.

### 8.2.4 Flow Battery

A flow battery is rechargeable in which an electrolyte is continuously supplied to one or more electrochemical cells. The electrolyte is stored and continuously supplied from tanks. It is simple to increase the amount of electrolyte held in the tanks of a simple flow battery to enhance the energy storage capacity. The power of the flow battery system can be adjusted by electrically connecting the electrochemical cells in series or parallel. A significant characteristic of flow battery systems is the decoupling of energy and power ratings.

The electrodes and membrane are the parts of the flow battery that uses polymers and nanomaterials to run the process.

Electrodes: The electrodes in flow batteries, which serve as the sites of the electrochemical reactions, are often made from nanomaterials. Nanomaterials with high surface areas, such as CNTs, graphene, and nanowires, can speed up chemical reactions in batteries. To stop the electrolyte from leaking out and to increase the electrode's stability and longevity, a polymer layer is applied to the electrodes. It was found that the boron-doped carbon felt enhanced electrochemical performances of vanadium redox flow batteries (VRFBs) (Figure 8.6). Kim et al. (2017) found that N and O co-doping through ammoxidation surface reactions with $NH_3$-$O_2$ successfully enhanced the electrocatalytic capabilities of graphite felt (GF) electrodes for VRFBs at high temperatures.

Chlorosulfonic acid was used with the aid of ultrasonication to etch the carbon paper used as the positive electrode for VRFBs. The facile treatment successfully produced carbon layer-exfoliated, wettability-enhanced, $SO_3H$-functionalized carbon paper. After the treatment, the electrochemical kinetics of the $VO^{2+}$/$VO^{2+}$ redox process on carbon paper greatly improved. Compared to untreated and only-soaking samples, carbon paper treated with ultrasonication shows improved electrochemical activity. The cell using treated carbon paper as the positive electrode exhibits a better discharge capacity and energy efficiency (He et al., 2018).

Membrane: The membrane in flow batteries that divides the two electrolyte solutions is made of polymers. The membrane is often built from materials that are chemically stable and can withstand deterioration from acidic or alkaline electrolytes, such as Nafion or polyethene oxide (PEO). Also, the polymer membrane aids in preventing cross-contamination between the two electrolyte solutions, which could impact the battery's performance. Hexyl tethered 1-methyl-2-mesityl-benzimidazolium ions functionalized polybenzimidazole through a hexyl linker

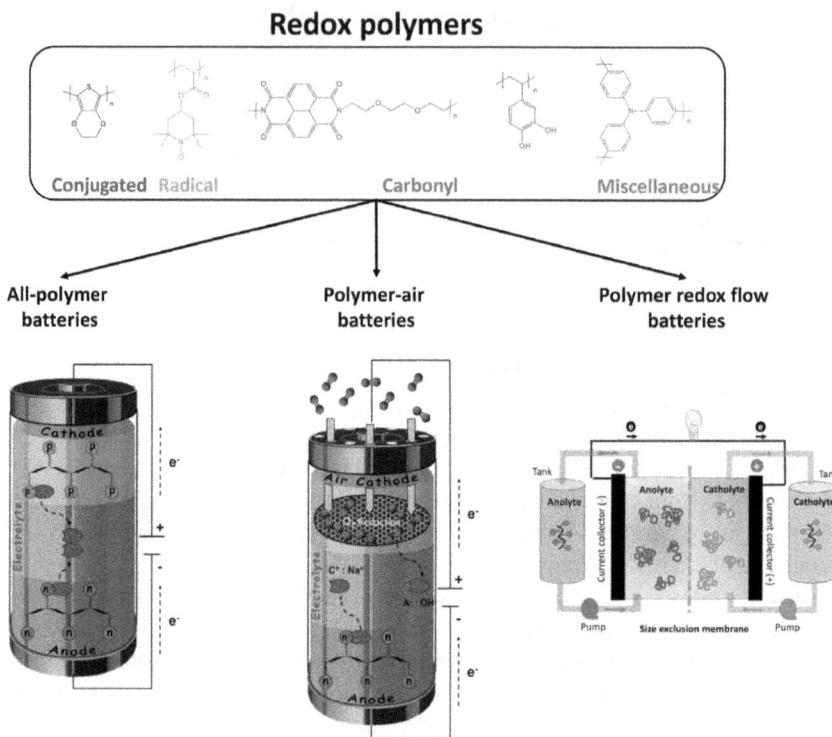

**FIGURE 8.6**
Various types of redox polymer-based batteries (Goujon et al., 2021). *Reproduced with permission from Elsevier.*

used as a membrane in vanadium flow batteries. The fabricated membrane showed better coulomb and voltage efficiency than polybenzimidazole (Lee et al., 2019).

## 8.3 Challenges of Using Functionalized Nanomaterials and Polymers in Energy Systems

Although functionalized materials have shown promising results in energy-related systems, there are still many factors that need to be overcome or worked on. Scalability is one of the main issues with the application of functionalized nanomaterials and polymers in energy systems. It can be challenging to scale up production to satisfy the demands of large-scale energy applications, despite the fact that many of these materials have demonstrated promising outcomes at the laboratory scale. The stability of these materials presents still another significant difficulty. Many polymers and functionalized nanomaterials can deteriorate over time due to exposure to radiation, high temperatures, or chemical processes. Its application in energy systems, where long-term stability is essential, may be constrained. Functionalized nanomaterials and polymers can be used in energy systems, but the cost of creating them can be limiting. Producing these materials can be costly, especially if specialist production is required. Safety concerns also exist about functionalized nanomaterials and polymers, particularly concerning their possible effects on the environment and human health. There is still much to learn about the possible risks associated with these materials, even though research has proven that many are safe when handled properly (F. Zhang, 2017).

## References

Branchi, M., Sgambetterra, M., Pettiti, I., Panero, S., & Navarra, M. A. (2015). Functionalized $Al_2O_3$ particles as additives in proton-conducting polymer electrolyte membranes for fuel cell applications. *International Journal of Hydrogen Energy, 40*(42), 14757–14767. https://doi.org/10.1016/j.ijhydene.2015.07.030

De Filpo, G., Nicoletta, F. P., Ciliberti, L., Formoso, P., & Chidichimo, G. (2015). Non-covalent functionalisation of single wall carbon nanotubes for efficient dye-sensitised solar cells. *Journal of Power Sources, 274*, 274–279. https://doi.org/10.1016/j.jpowsour.2014.10.053

Gao, J., Dong, X., Tian, Q., & He, Y. (2023). Carbon nanotubes reinforced proton exchange membranes in fuel cells: An overview. *International Journal of Hydrogen Energy, 48*(8), 3216–3231. https://doi.org/10.1016/j.ijhydene.2022.10.173

Gong, J., Liang, J., & Sumathy, K. (2012). Review on dye-sensitized solar cells (DSSCs): Fundamental concepts and novel materials. *Renewable and Sustainable Energy Reviews, 16*(8), 5848–5860. https://doi.org/10.1016/j.rser.2012.04.044

Gouda, M. H., Elessawy, N. A., & Toghan, A. (2021). Development of effectively costed and performant novel cation exchange ceramic nanocomposite membrane based sulfonated PVA for direct borohydride fuel cells. *Journal of Industrial and Engineering Chemistry, 100*, 212–219. https://doi.org/10.1016/j.jiec.2021.05.021

Goujon, N., Casado, N., Patil, N., Marcilla, R., & Mecerreyes, D. (2021). Organic batteries based on just redox polymers. *Progress in Polymer Science, 122*, 101449. https://doi.org/10.1016/j.progpolymsci.2021.101449

Halankar, K. K., Mandal, B. P., & Tyagi, A. K. (2021). Superior electrochemical performance of MoS2 decorated on functionalized carbon nanotubes as anode material for sodium ion battery. *Carbon Trends, 5*, 100103. https://doi.org/10.1016/j.cartre.2021.100103

He, Z., Jiang, Y., Li, Y., Zhu, J., Zhou, H., Meng, W., Wang, L., & Dai, L. (2018). Carbon layer-exfoliated, wettability-enhanced, $SO_3H$-functionalized carbon paper: A superior positive electrode for vanadium redox flow battery. *Carbon, 127*, 297–304. https://doi.org/10.1016/j.carbon.2017.11.006

Hussain, I., Tran, H. P., Jaksik, J., Moore, J., Islam, N., & Uddin, M. J. (2018). Functional materials, device architecture, and flexibility of perovskite solar cell. *Emergent Materials, 1*(3), 133–154. https://doi.org/10.1007/s42247-018-0013-1

Jha, N., Ramesh, P., Bekyarova, E., Tian, X., Wang, F., Itkis, M. E., & Haddon, R. C. (2013). Functionalized single-walled carbon nanotube-based fuel cell benchmarked against US DOE 2017 technical targets. *Scientific Reports, 3*(1), Article 1. https://doi.org/10.1038/srep02257

Kadem, B., Hassan, A., Göksel, M., Basova, T., Şenocak, A., Demirbaş, E., & Durmuş, M. (2016). High performance ternary solar cells based on P3HT:PCBM and ZnPc-hybrids. *RSC Advances*, *6*(96), 93453–93462. https://doi.org/10.1039/C6RA17590B

Kim, J., Lim, H., Jyoung, J.-Y., Lee, E.-S., Yi, J. S., & Lee, D. (2017). High electrocatalytic performance of N and O atomic co-functionalized carbon electrodes for vanadium redox flow battery. *Carbon*, *111*, 592–601. https://doi.org/10.1016/j.carbon.2016.10.043

Kishore Kumar, D., Kříž, J., Bennett, N., Chen, B., Upadhayaya, H., Reddy, K. R., & Sadhu, V. (2020). Functionalized metal oxide nanoparticles for efficient dye-sensitized solar cells (DSSCs): A review. *Materials Science for Energy Technologies*, *3*, 472–481. https://doi.org/10.1016/j.mset.2020.03.003

Lee, Y., Kim, S., Maljusch, A., Conradi, O., Kim, H.-J., Jang, J. H., Han, J., Kim, J., & Henkensmeier, D. (2019). Polybenzimidazole membranes functionalised with 1-methyl-2-mesitylbenzimidazolium ions via a hexyl linker for use in vanadium flow batteries. *Polymer*, *174*, 210–217. https://doi.org/10.1016/j.polymer.2019.04.048

Liew, K. B., Wan Daud, W. R., Ghasemi, M., Loh, K. S., Ismail, M., Lim, S. S., & Leong, J. X. (2015). Manganese oxide/functionalised carbon nanotubes nanocomposite as catalyst for oxygen reduction reaction in microbial fuel cell. *International Journal of Hydrogen Energy*, *40*(35), 11625–11632. https://doi.org/10.1016/j.ijhydene.2015.04.030

Ma, C., Zhang, J., Xu, M., Xia, Q., Liu, J., Zhao, S., Chen, L., Pan, A., Ivey, D. G., & Wei, W. (2016). Cross-linked branching nanohybrid polymer electrolyte with monodispersed $TiO_2$ nanoparticles for high performance lithium-ion batteries. *Journal of Power Sources*, *317*, 103–111. https://doi.org/10.1016/j.jpowsour.2016.03.097

Makvandi, P., Iftekhar, S., Pizzetti, F., Zarepour, A., Zare, E. N., Ashrafizadeh, M., Agarwal, T., Padil, V. V. T., Mohammadinejad, R., Sillanpaa, M., Maiti, T. K., Perale, G., Zarrabi, A., & Rossi, F. (2021). Functionalization of polymers and nanomaterials for water treatment, food packaging, textile and biomedical applications: A review. *Environmental Chemistry Letters*, *19*(1), 583–611. https://doi.org/10.1007/s10311-020-01089-4

Mishler, J., Wang, Y., Mukundan, R., Spendelow, J., Hussey, D. S., Jacobson, D. L., & Borup, R. L. (2012). Probing the water content in polymer electrolyte fuel cells using neutron radiography. *Electrochimica Acta*, *75*, 1–10. https://doi.org/10.1016/j.electacta.2012.04.040

Nelson, J. (2011). Polymer: fullerene bulk heterojunction solar cells. *Materials Today*, *14*(10), 462–470. https://doi.org/10.1016/S1369-7021(11)70210-3

Niederhausen, J., Mazzio, K. A., & MacQueen, R. W. (2021). Inorganic–organic interfaces in hybrid solar cells. *Electronic Structure*, *3*(3), 033002. https://doi.org/10.1088/2516-1075/ac23a3

Raduwan, N. F., Shaari, N., Kamarudin, S. K., Masdar, M. S., & Yunus, R. M. (2022). An overview of nanomaterials in fuel cells: Synthesis method and application. *International Journal of Hydrogen Energy*, *47*(42), 18468–18495. https://doi.org/10.1016/j.ijhydene.2022.03.035

Su, H., & Hu, Y. H. (2021). Recent advances in graphene-based materials for fuel cell applications. *Energy Science & Engineering*, *9*(7), 958–983. https://doi.org/10.1002/ese3.833

Wong, W. Y., Daud, W. R. W., Mohamad, A. B., Kadhum, A. A. H., Majlan, E. H., & Loh, K. S. (2012). Nitrogen-containing carbon nanotubes as cathodic catalysts for proton exchange membrane fuel cells. *Diamond and Related Materials*, *22*, 12–22. https://doi.org/10.1016/j.diamond.2011.11.004

Xu, P., Chang, X., Liu, R., Wang, L., Li, X., Zhang, X., Yang, X., Wang, D., & Lü, W. (2020). Boosting power conversion efficiency of quantum dot-sensitized solar cells by integrating concentrating photovoltaic concept with double photoanodes. *Nanoscale Research Letters*, *15*(1), 188. https://doi.org/10.1186/s11671-020-03424-8

Zhang, F. (2017). Grand challenges for nanoscience and nanotechnology in energy and health. *Frontiers in Chemistry*, *5*. www.frontiersin.org/articles/10.3389/fchem.2017.00080

Zhang, W., Chen, J., Swiegers, G. F., Ma, Z.-F., & Wallace, G. (2010). Microwave-assisted synthesis of Pt/CNT nanocomposite electrocatalysts for PEM fuel cells. *Nanoscale*, *2*(2), 282–286. https://doi.org/10.1039/B9NR00140A

# 9

## Potentials and Challenges of Functional Nanomaterials and Polymers

### 9.1 Brief Discussion on Potential Industrial Application of Functional Polymers and Nanomaterials

According to our assessment, nanotechnology is a broadly enabling technology with the potential to significantly and permanently boost a wide range of industry sectors. Improved human health diagnostic systems and better ways to detect, prevent, or treat the adverse effects of chemical and biological agents are a few examples of smarter, more rapid, and practical communication components and devices; smart materials that respond to external stimuli; cleaner energy; safer, efficient, and environmentally friendly manufacturing processes; and improved manufacturing processes. Global societal benefits will arise from the commercialization of nanotechnology, boosting productivity, improving health, and improving quality of life.

Nanotechnology will not be dominated by any individual, organization, sector, or nation. According to Qiu Zhao et al. (2003) and Roco (2011), the United States does not hold a commanding advantage in government or private industrial nanotechnology investment. Since 1997, governments of most industrialized countries have invested in nanotechnology, with prominent ongoing programs in the United States, Japan, and Western Europe, as well as in Canada, Eastern Europe, Australia, China, Israel, Singapore, Korea, and Taiwan. This is fundamentally in contrast to the prior industrial revolution, which provided rise to a small number of dominant countries while the rest of the world was still in development.

In the Nineteenth century, for instance, DuPont was engaged in nanomaterials research. As a great materials pioneer with a long history of leadership in safeguarding the health, safety, and well-being of employees, clients, and society at large, we are carrying on our legacy. The change history of the DuPont company is Established in 1802, DuPont produced explosives and gunpowder for most of the nineteenth century. One of the first industrial research laboratories was created there in 1903. Most of the second wave's profitable items were discovered and developed at the Experimental Station in Wilmington, DE. These are well-known industrial polymers that have enhanced the lives of people all over the world, such as nylon, Teflon, polyester, Lycra, Kevlar, Nomex, Tyvek, and many others. DuPont is investing in biotechnology and nanotechnology because it has a strong commitment to material innovation and wants to change industries in the next 100 years. In order to provide a strong growth rate based on fresh and cutting-edge products from each

DOI: 10.12019781003391364-9

of these platforms, DuPont has divided its companies into five growth platforms. Nanotechnology will play an increasingly significant role in this shift as we discover new properties originating from the nanoscale and learn how to incorporate them into products with new, enhanced functionality that customers seek.

## 9.2 Chemical Engineering Industries: Nanomaterial Fabrication Techniques and Challenges

Advancements in chemical engineering unit operations in the early 1900s were crucial to the successful commercialization of a number of chemical industry innovations. Understanding and regulating nanotechnology unit operations will be equally critical for nanotechnology commercialization. Nanotechnology platforms and process technologies are used in our engineering nanotechnology laboratory. The platforms involve nanoscale synthesis, nanolayer coatings and encapsulation, particle design, and dispersion science to integrate the building blocks into final product design.

## 9.3 Nanoscale Synthesis of Carbon-based Nanoparticles

Nanoparticles can be synthesized by milling huge particles or by directed chemical synthesis. Particles as fine as 100 nm can be broken down or dispersed using microfluidizers and small media mills. Carbon nanotubes (CNTs) are created from liquid or vapor phases, like other nanoparticles. Metal, metal oxide, and ceramic nanoparticles can be produced on a large scale via chemical and physical vapor phase synthesis. Having been used for many years, carbon black, color pigments, and fumed silica are among the most well-known commercial nanoparticles. High-temperature flames, thermal plasma, and lasers are all potential heat sources for the vapor phase synthesis process. Spray pyrolysis is a proven method for developing complex nanoparticles. Controlling the particle size distribution (PSD), *in situ* production of hybrid particles and structures, effective collecting systems, and continuous operations at high throughput and productivity would be challenges for vapor phase synthesis of nanoparticles. Nanoparticles use conventional processes such as precipitation, sol-gel, crystallization, and emulsion polymerization. Molecular template nucleation can alter crystal size, position, and form. Technically difficult tasks include classifying nanoparticle sizes and separating nanoparticles in solid–liquid and solid–gas systems. A novel biocatalysis method that we are developing is being tested on a small scale at a pilot plant. CNT output has been published in a number of ways. It appears that the trend is moving away from laser and plasma technologies and toward chemical vapor deposition (CVD) strategies. The availability of quantity and diversity, as well as constant product quality, continue to be rate-limiting steps for many real-world applications. Despite their huge potential and considerable interest, CNTs are still in the research phase. Recently, employing concepts from biotechnology, the separation of conducting tubes from semiconducting tubes was demonstrated in various research labs, as previously reported (Zheng et al., 2003).

## 9.4 TiO$_2$ Particle Dispersion and Coatings

Surface atoms constitute 20% of particles with a diameter of 10 nm, 80% of particles with a diameter of 2 nm, and 100% of particles with a diameter of 1 nm. A single-walled CNT with a diameter of 1 nm contains every carbon atom on its surface. Small, high surface-area nanoparticles differ from bulk materials in terms of their optical, chemical, and physical characteristics. While TiO$_2$ pigment particles of size 25 nm are transparent in the visible range but effectively block UV light, TiO$_2$ pigment particles of size 250 nm are perfect for hiding power by dispersing visible light. In addition, TiO$_2$ nanoparticles are more photoactive than micron-sized pigment particles. Controlling the surface coating of nanoparticles is essential since it has been found that metal oxide nanoparticles have greater chemisorption potential than simple surface area-based adsorption. A broad, dense coating or encapsulation is needed to passivate optical or chemical activities. Surface coatings are crucial for avoiding, dispersing, and stabilizing agglomerates. Compatible coatings must be integrated into host matrices to rapidly obtain the benefits of nanoparticles. Both dry and wet methods can be used to apply coatings. However, nanoparticles is a challenging operation. A high-quality coating needs a good dispersion, but dispersion of nanoparticles, whether dry or wet, is difficult. For achieving distinctive features in nanocomposite applications, stabilizing dispersion and controlling nanoparticle surface coating are still needs that must be satisfied. Both academic and commercial researchers may find these to be important research areas. Self-assembly is favorable due to the multiplicity of interparticle forces and the difficulties in dispersion and coating of nanoparticles. Molecular templates and biomolecule-assisted assembly are examples of fluid dynamic-driven self-assembly techniques. Other techniques include electric or magnetic field-induced assembly, injection, nanojets, microfluidization, or micronization devices. The potential for growth of the last of these prospects is tremendous (Qiu Zhao et al., 2003).

## 9.5 Overview of Functional Materials Used in Biofuel Industries

Functional materials are used in various processes due to the versatility in the functions they can perform. Functional materials have few properties that make them function on their own, e.g., electromagnetic materials and piezoelectricity. They show excellence in magnetic, catalytic, electrical, optical, and mechanical properties. Climate change and environmental pollution are two of the most severe global concerns confronting humanity today. Developing various functional materials that can address many of the issues associated with both present and future tactics is an important component of such research. For instance, creating materials with catalytic properties to convert renewable resources is possible. Biofuels can be produced from biomass and then used as an alternative energy source.

Functional materials have been made by modifying the nanoparticles or layers of carbon polymers by incorporating a new element in between, thus enhancing the characteristic features of the material. A wide range of applications include desalination, fuel cells and batteries, and bioremediation such as the removal of heavy metals, pesticides, and antibiotics.

---

## 9.6 Commonly Used Functional Materials and Processes

There are various base materials being used. They include organic and inorganic materials. The inorganic materials are primarily oxides of titanium, silicon, aluminum, zinc, ferrous, etc. For example,

i. $TiO_2$: Titanium dioxide can be converted into $TiO_2$-$NH_2$ or $TiO_2$-$COOH$. This is mainly due to the availability of vacant d and f orbitals in the inorganic elements, which allows them to combine with other compounds and enhance its functional properties. The $TiO_2$ nanoparticles have an extremely large surface area/particle size and thus can also form agglomerates;

ii. The other set of materials are compounds of carbon (organic materials) such as CNTs (single-walled and multiwalled) and reduced graphene oxide (rGO);

iii. Sometimes even membrane materials such as polyethylene sulfone (PES), polyvinyl fluoride (PVDF), and polyethylenimine (PEI) undergo either sulfonation or amination process to reform into functional materials;

iv. Bimetallic nanomaterials;

v. Nanomaterials with a mesoporous ceramic structure;

vi. Polymeric nanomaterials;

vii. Bio-nanomaterials;

viii. Metal–organic frameworks (MOFs);

ix. Core–shell nanomaterials;

x. Nanocomposites: i.e., sub-micrometric mixtures (1–100 nm) of materials of a similar nature;

xi. Composites: i.e., mixtures of materials consisting of a matrix with micrometric dispersion;

xii. Hybrids: i.e., sub-micrometric mixtures of materials of a different nature compared to the compound hybrid material;

xiii. Nanohybrid.

**TABLE 9.1**

Various additives used for functionalization of nanoparticles

| Nanoparticle | Additive | Functional Material |
|---|---|---|
| $TiO_2$ | $NH_2$ | $TiO_2\text{-}NH_2$ |
| $SiO_2$ | $NH_2$ | $SiO_2\text{-}NH_2$ |
| $TiO_2$ | COOH | $TiO_2\text{-}COOH$ |
| PES | $NH_2$ | PES-$NH_2$ |
| PVDF | Sulfonation | S-PVDF |
| PEI | Sulfonation | S-PEI |
| C60 | $La_2O_3$ | $La_2O_3\text{-}C60$ |

## 9.7 Application of Functional Nanomaterials and Polymers in Food and Textile Industries

### 9.7.1 Nanotechnology and Functionalized Materials in Food Industries

The vast majority of nanotechnology research is devoted to the development of bio-science and engineering applications. The methods for using nanotechnology in the food industry are very different from those used in more traditional applications (Figure 9.1). Food processing is a multi-technology manufacturing company that uses a wide variety of raw materials, severe biosafety regulations, and strictly monitored technical procedures. Creating new functional materials, micro- and nanoscale processing, product development, and developing techniques and instrumentation for enhanced food safety and biosecurity are the four main areas of food production that nanotechnology has the potential to improve. The effect of food material characteristics on nutritional value and bioavailability at the nanoscale level has been emphasized. Additionally, the association between bulk physicochemical properties and food material shape has been investigated (Weiss et al., 2006).

#### 9.7.1.1 Usage of Functionalized Polylactic Acid in Food Industry

Polylactic acid (PLA), first created in 1932, is a fundamental component of several biodegradable nanoparticles. Nevertheless, it was apparently inappropriate for use in biomedical or agricultural applications due to its high cost and susceptibility to hydrolytic breakdown; therefore, it was only employed sparingly in research. However, PLA's potential as a suture material was identified in the 1970s, and in the 1980s, a method for making the polymer by bacterial fermentation was devised, which significantly boosted its yield and decreased prices (Lunt, 1998). Numerous manufacturers now offer PLA in large quantities. In biomedicine, biodegradable polymers are used for a variety of purposes, such as decreasing the need for surgical implantation and removal and encapsulating and conveying a variety of molecules (such as drugs, vaccinations, and proteins). Due to its

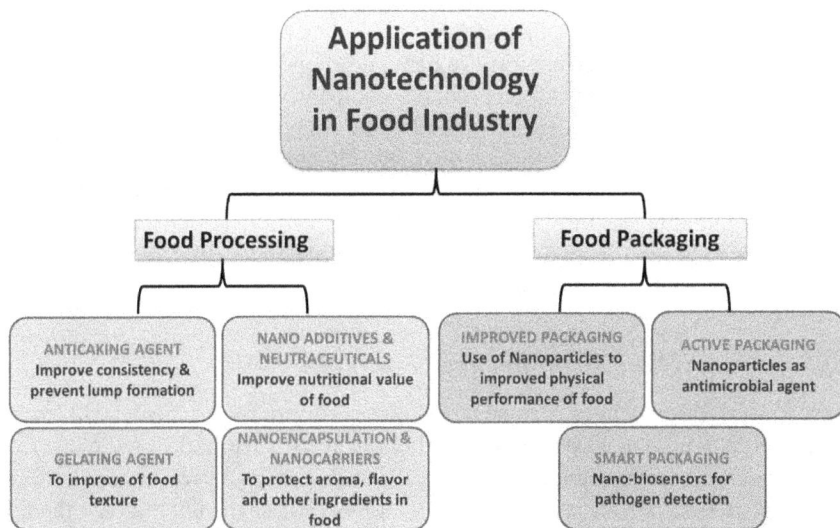

**FIGURE 9.1**
Application of nanomaterials and polymers in food industries (T. Singh et al., 2017). *Reproduced with permission from Frontiers.*

powerful encapsulating characteristics, PLA is one of the fundamental building blocks of many biodegradable nanoparticles, although it also has drawbacks. PLA is quickly eliminated from the blood and retained in the liver and kidneys for further use. While this is the best way to treat intracellular pathogens isolated from these sites, it is less suitable for getting active ingredients to other body parts. Additionally, PLA breaks down in the intestinal fluid, which reduces its usefulness as an oral delivery vehicle. Combining the hydrophobic PLA nanoparticle with a hydrophilic substance like polyethylene glycol (PEG) helps resolve these problems. A drug to be supplied can be contained in a micellar-like assembly of nanoparticles formed of a PLA-PEG diblock copolymer.

PEG also influences particle zeta potential; that is, it has a fewer negative surface charge (approximately −6 mV for 30:5 PLA:PEG copolymer and approximately −50 mV for PLA alone). According to Tobío et al. (2000), investigated the PEG coating on PLA nanoparticles, its interaction and impact with biological surfaces. PEG capping of the carboxy PLA end groups or from displacing the shear plane of the diffusion layers away from the nanoparticle. Changes in zeta potential are important because they alter particle interactions with other chemicals in the food system. The surface charge of a particle is crucial if it is to be used as a building block for more intricate structures like nanolaminates. Finally, depending on the base biopolymer employed to create the nanoparticles, particle surfaces might either be hydrophobic or hydrophilic (Weiss et al., 2006).

### 9.7.1.2 Food Processing

Food additives with nanostructures are being developed to enhance flavor, consistency, and texture. Numerous food products' shelf-lives were extended using nanotechnology, contributing to decreased food waste from microbial contamination. In order to deliver food additives to food products without changing their fundamental form, nanocarriers are currently being used. The transport of any bioactive substance to different places throughout the body may be directly impacted by particle size because only submicron nanoparticles may be absorbed efficiently in some cell lines but not bigger size micro-particles.

### 9.7.1.3 Preservation, Shelf-life, and Nutritional Value

These bioactive compounds may be easily absorbed into food items when they are encapsulated, which makes it easier for them to resist such harsh conditions than when they are not encapsulated. This is because non-capsulated versions of these compounds have poor water solubility, which makes it challenging to use them in food products. In order to improve the delivery of medications, vitamins, or delicate micronutrients in daily meals, nanoparticle-based microscopic edible capsules have been developed. In functional foods, where bioactive ingredients frequently deteriorate and ultimately result in inactivation because of the hostile environment, nanoencapsulation of these bioactive ingredients prolongs the shelf-life of food products by slowing or preventing degradation until the product reaches the target site (Oppenheim, 1981). Additionally, edible nano-coatings on various food components could offer flavors, colors, antioxidants, enzymes, and anti-browning agents in addition to serving as a barrier to moisture and gas exchange. Even after the package is opened, they might still help manufactured foods last longer. Encapsulating functional components into droplets can often slow down chemical breakdown processes by modifying the properties of the interfacial layer surrounding them (Ibrahim et al., 1992). For instance, curcumin, the most active and least stable bioactive component of turmeric (*Curcuma longa*) showed decreased antioxidant activity and was discovered to be stable to pasteurization and of variable ionic strength after encapsulation (Weiss et al., 2006).

### 9.7.1.4 Food Packaging

Nano-based "smart" and "active" food packaging has several advantages over traditional packaging methods, including better packaging material with improved mechanical strength, barrier properties, and antimicrobial films, as well as nanosensing for pathogen detection and alerting consumers to the safety status of food. Low quantities of inorganic nanoparticles exhibit significant antibacterial activity and are more stable under challenging conditions. An example of active packaging that inhibits or delays the growth of microbes on the food surface is antimicrobial packaging, which comes into contact with the food product or the headspace inside. Numerous nanoparticles, including silver, copper, chitosan, and metal oxide nanoparticles like titanium or zinc oxide, have exhibited antibacterial features. The polymer matrix can be made lighter, stronger, and more fire resistant and have superior thermal properties by adding inert nanoscale fillers like chitin or chitosan, silica ($SiO_2$) nanoparticles, or clay and silicate nanoplatelets. Antimicrobial nanocomposite films made by impregnating fillers (with at least one dimension in the nanometric range or nanoparticles) with polymers have a dual advantage due to their structural integrity and barrier properties.

## 9.7.2 Textile Industries

In the past, individuals used durability as their primary criterion when choosing a fabric. Later, they turned to aesthetic considerations and comfort levels as their main deciding factors. The way that consumers now approach clothing and textiles is changing, and they are looking for some added usefulness in addition to classic textile characteristics. Protective apparel, temperature-regulating, industrial, sports, and automobile textiles can all contribute to functionality. To qualify as functional textiles, all of the aforementioned textile materials must maintain at least one particular utility. Textiles with added capabilities of modifying and regulating various characteristics, such as temperature, humidity, color, and the controlled release of some additives from fibers, are known as functional materials in the textile industry. Polyester and viscose are the most widely utilized fibers in the production of many types of practical textiles.

### 9.7.2.1 Textile Functional Materials

9.7.2.1.1 Phase-change Materials, Fire Retardants, and Fragrance Finishes

Micro- and nanoencapsulation of phase change materials may convert the core material from solid to liquid and liquid to stable by altering the entropy within a specific temperature range. This method makes the targeted subject's exposure to temperature change lessened. To maintain uniform clothing temperatures, phase transition materials are encapsulated. Microencapsulated phase change materials enhance the comfort delivery in blankets, duvets, mattresses, snowsuits, and vests.

Although fragrance finishes are directly applied to textiles, aroma stability only lasts for a limited number of wash cycles. The fragrant functioning of the cloth is extended for a much longer period of time using fragrance micro- and nanoencapsulation. For aromatherapy to treat headaches, insomnia, and unpleasant odor, this method is mostly used to encapsulate essential oil fragrances such as pine, rosemary, and lavender.

The loss of softness and other low-stress mechanical properties that are seeded by the direct application of flame-retardant chemicals is not allowed by the encapsulation of flame-retardant materials. Organophosphorus compounds as the core and nitrogenous compounds as the sheath can have a synergistic effect when flame-retardant materials are chosen scientifically. Some intumescent flame-retardant coatings can also be made using micro- and nanoencapsulation processes. This flame retardancy technique is commonly employed in the military to treat tentage, furniture, and firefighting gear (M. K. Singh, 2021).

### 9.7.2.1.2 Functionality in Textiles—Microencapsulation

Microencapsulation is a method in which an active component is contained in a small area and coated or covered with a thin polymeric layer to protect the core substance while allowing for controlled release. Diverse shell wall materials, including natural polymers (gelatine, cellulose, chitosan, etc.), artificial polymers (cellulosic derivatives), and synthetic polymers (polyamide, polyester, etc.), can be used to encapsulate various compounds, including dyes, proteins, fragrance, monomers, and catalysts. The loading content can range from 5% to 90% of the weight of the microparticles. Biocides, insecticides, essential oils, moisturizers, energizers, medicinal oils, and vitamin E may now be uploaded into fabrics via micro- and nanoencapsulation (Nelson, 2002).

### 9.7.2.1.3 Deodorant Functional Textiles

Deodorant additives are added during fiber extrusion, deodorant finishes are applied to the fiber surface after spinning, and the deodorant fibers are produced by altering the polymer molecular chain during polymerization. The simplest way to impart deodorants in fibers is to apply a deodorant additive as a dope additive. A photocatalyst was used to modify the polyester staple fibers, which were then combined with cotton and bamboo fibers to create a variety of fabrics. Different samples had photocatalyst fiber contents that ranged from 0% to 100%, with 20% increments. We investigated and evaluated the cloth samples' deodorant potential (Figure 9.2). It is stated that a fabric's deodorant potential improves when the photocatalyst fiber content increases to 80% or 100% but that the photocatalyst content in fabric samples is still between 40% and 60%, resulting in a minimal deodorizing effect. It has been determined that a photocatalyst content of at least 80% is necessary to create a usable deodorant fabric (Zhang et al., 2012).

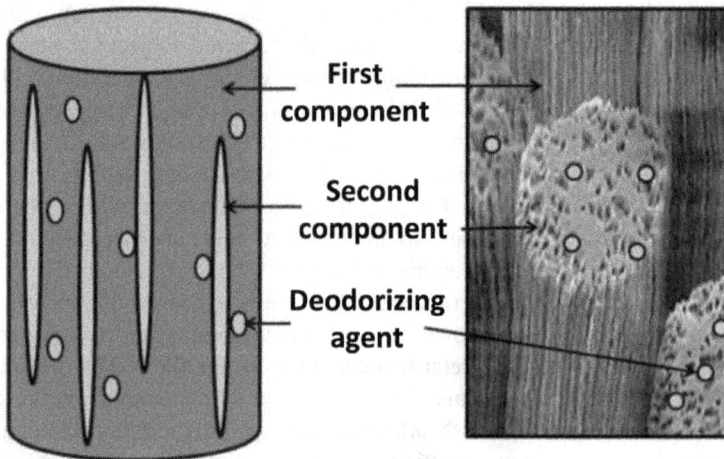

**FIGURE 9.2**
Inherent deodorant textile fiber using nanomaterials (M. K. Singh, 2021). *Reproduced with permission from IntechOpen.*

*9.7.2.1.4 Benefits of Functionalization by Microencapsulation*

i. Since the majority of the active ingredients are flammable, chemically brittle, or physically, chemically, or thermally unstable, they must be encased in a protective shell before being applied directly to textile substrates. The micro- or nanoencapsulation not only protects the active substances from environmental stimuli (acidity, alkalinity, heat, moisture, and oxidation) but also limits their interactions with other compounds that are still in the system, to lengthen the time before functionality is improved.

ii. The capsules can also inhibit volatile chemicals from evaporating.

iii. In addition, the microcapsules protect both users and manufacturers from exposure to dangerous material. Before processing, active compounds can be handled safely, allowing microcapsules to enable soluble substances to undergo temporary insoluble transformation. Using this method, the unpleasant aroma produced by active chemicals can be masked prior to final application during manufacture (M. K. Singh, 2021).

## 9.8 Industrial Challenges of Functional Materials

There are a few obstacles to the incomparable influence of nanotechnology on the sector, notwithstanding its projected results. The absence of information, the potential for adverse consequences on the environment, human welfare, security, and sustainability are still problems (Santos et al., 2015). Turning beyond the bench is hindered by a number of factors, including a lack of funding for applied research, financial backers' concern when dealing with new developments, and a lack of government direction for the administrative management of nanotechnology. The possible toxicity of the nanomaterials themselves is another area of worry. Numerous studies on the toxicity of nanomaterials have been conducted in response to this worry, and the results suggest that occupational exposure to nanoparticles may be counterproductive. However, if the nanomaterials are needed for a composite material, the impact of introducing them into the environment is more complex. Once in the environment, processes like aggregation; adsorption of normal, natural matter; sedimentation; and oxidation change the nanomaterials from their pristine structure and can modify their toxicity. Numerous applications of the nanomaterials and nanotechnologies described here are not readily available in the marketplace. By lowering treatment costs and, astonishingly, making it possible to cure contaminants that are currently untreatable, the application of nanotechnology in existing water treatment has the potential to alter many of these cycles. The contemporary water treatment industry has distinct needs, and it would be advantageous for the entire sector to work together with analysts and end-users to discover answers to well-defined issues. The following are issues with current industrial wastewater treatment methods: In order to successfully treat contaminated water, it is necessary to safeguard both the downstream unit processes and the long-term production of suitable water quality. Industrial wastewater treatment is challenging and typically requires a long series of successive treatment steps, or a "treatment train." Even if distillation of contaminated water is guaranteed to remove all contaminants except volatile ones, the cost can be too high. Additionally, not all industrial water requires such extreme treatment. For instance, as long as specific impurities are removed, water produced for internal reuse in the oil and gas industry does not need to be completely desalinated. Therefore, the economic viability of many industrial activities would be greatly improved by a targeted treatment technique (rather than brute-force treatment such as distillation) that removes just specific contaminants and generates tailored water quality. In addition, a lot of the therapeutic methods used today are ineffective and problematic. For instance, oil in wastewater may delaminate membranes, alter their capacity for separation, and impair the operation of various adsorbers, requiring an oil–water separation step before further treatment is possible. Likewise, microorganisms evading pretreatment will foul desalination membrane surfaces, but unquenched disinfectants in pretreated water may oxidize or destroy the membranes. Reverse osmosis membranes are limited by osmotic and hydraulic pressure as well as mineral precipitation; adsorbers are restricted to only certain compounds with specific functional groups or structures; and most ion exchange surfaces are chemical-specific. Furthermore, some technologies have inherent limitations, particularly at high water recovery and salinity. Unit processes are needed to simultaneously remove contaminants from various chemical groups (organic, inorganic, and particulate matter), recover high-value materials (like rare earth elements, nutrients, acids/bases, and energy), and do this from various industrial waste streams (Jassby et al., 2018). A significant hypothetical advantage of nanotechnology examination is determining whether existing hypotheses are useful indicators of what will occur in enterprises where nanotechnology applications are being developed, as well as changing and expanding existing hypotheses as justified by research discoveries. Then, speculations could be made to thoroughly examine hypothetical structures that already exist. Competitors turn to item highlights and cycle efficiencies when growth is slow. Engaging in activities that encourage continuous improvement, including factual interaction control, measure advancement, preventative maintenance projects, and value design, is crucial. Effectiveness is typically supported by the robotic traits of normalization, specialization, and centralization. Client demands and suggestions drive the statistical surveying trend (Shea, 2005). Finally, worldwide nanomaterial regulation is critical for the safe use of new, large nanomaterial technologies all over the world (Subhan et al., 2021).

---

# References

Ibrahim, H., Bindschaedler, C., Doelker, E., Buri, P., & Gurny, R. (1992). Aqueous nano-dispersions prepared by a salting-out process. *International Journal of Pharmaceutics*, *87*(1), 239–246. https://doi.org/10.1016/0378-5173(92)90248-Z

Jassby, D., Cath, T. Y., & Buisson, H. (2018). The role of nanotechnology in industrial water treatment. *Nature Nanotechnology*, *13*(8), Article 8. https://doi.org/10.1038/s41565-018-0234-8

Lunt, J. (1998). Large-scale production, properties and commercial applications of polylactic acid polymers. *Polymer Degradation and Stability*, *59*(1), 145–152. https://doi.org/10.1016/S0141-3910(97)00148-1

Nelson, G. (2002). Application of microencapsulation in textiles. *International Journal of Pharmaceutics*, *242*(1), 55–62. https://doi.org/10.1016/S0378-5173(02)00141-2

Oppenheim, R. C. (1981). Solid colloidal drug delivery systems: Nanoparticles. *International Journal of Pharmaceutics*, *8*(3), 217–234. https://doi.org/10.1016/0378-5173(81)90100-9

Qiu Zhao, Q., Boxman, A., & Chowdhry, U. (2003). Nanotechnology in the chemical industry—opportunities and challenges. *Journal of Nanoparticle Research*, *5*(5), 567–572. https://doi.org/10.1023/B:NANO.0000006151.03088.cb

Roco, M. C. (2011). The long view of nanotechnology development: The national nanotechnology initiative at 10 years. *Journal of Nanoparticle Research*, *13*(2), 427–445. https://doi.org/10.1007/s11051-010-0192-z

Santos, C. S. C., Gabriel, B., Blanchy, M., Menes, O., García, D., Blanco, M., Arconada, N., & Neto, V. (2015). Industrial applications of nanoparticles—a prospective overview. *Materials Today: Proceedings*, *2*(1), 456–465. https://doi.org/10.1016/j.matpr.2015.04.056

Shea, C. M. (2005). Future management research directions in nanotechnology: A case study. *Journal of Engineering and Technology Management*, *22*(3), 185–200. https://doi.org/10.1016/j.jengtecman.2005.06.002

Singh, M. K. (2021). Textiles functionalization—A review of materials, processes, and assessment. In *Textiles for functional applications*. IntechOpen. https://doi.org/10.5772/intechopen.96936

Singh, T., Shukla, S., Kumar, P., Wahla, V., Bajpai, V. K., & Rather, I. A. (2017). Application of nanotechnology in food science: Perception and overview. *Frontiers in Microbiology*, *8*. www.frontiersin.org/articles/10.3389/fmicb.2017.01501

Subhan, M. A., Choudhury, K. P., & Neogi, N. (2021). Advances with molecular nanomaterials in industrial manufacturing applications. *Nanomanufacturing*, *1*(2), Article 2. https://doi.org/10.3390/nanomanufacturing1020008

Tobío, M., Sánchez, A., Vila, A., Soriano, I., Evora, C., Vila-Jato, J. L., & Alonso, M. J. (2000). The role of PEG on the stability in digestive fluids and in vivo fate of PEG-PLA nanoparticles following oral administration. *Colloids and Surfaces. B, Biointerfaces*, *18*(3–4), 315–323. https://doi.org/10.1016/s0927-7765(99)00157-5

Weiss, J., Takhistov, P., & McClements, D. J. (2006). Functional materials in food nanotechnology. *Journal of Food Science*, *71*(9), R107–R116. https://doi.org/10.1111/j.1750-3841.2006.00195.x

Zhang, H., Ge, C., Zhu, C., Li, Y., Tian, W., Cheng, D., & Pan, Z. (2012). Deodorizing properties of photocatalyst textiles and its effect analysis. *Physics Procedia*, *25*, 240–244. https://doi.org/10.1016/j.phpro.2012.03.078

Zheng, M., Jagota, A., Semke, E. D., Diner, B. A., Mclean, R. S., Lustig, S. R., Richardson, R. E., & Tassi, N. G. (2003). DNA-assisted dispersion and separation of carbon nanotubes. *Nature Materials*, *2*(5), Article 5. https://doi.org/10.1038/nmat877

# 10

## Conclusion: Functional Nanomaterials and Polymers

Nanomaterials have a long history, and people have inadvertently used them. Feynman's famous address "There's plenty of room at the bottom" introduced the concept of modern nanotechnology to academia. Following this, there has been significant progress in nanotechnology, and the subject is constantly expanding into new fields. Materials are called nanoparticles (NPs) when their dimensions fall between 1 and 100 nm (Baig et al., 2021).

Numerous complex smart materials made possible by nanotechnology hold great potential for helping solve some of the world's present problems in the areas of contemporary energy generation, wastewater treatment, desalination, and pollution detection and degradation. Multifunctional nanomaterials have the potential to be more sensitive materials and new, sophisticated uses for bioremediation, energy storage, and industrial wastewater treatment. Smart nanomaterials can be used to create new, lightweight batteries with high power densities in addition to producing sustainable energy from renewable resources. Nanomaterials can also aid in solving environmental problems including recycling waste and improving the quality of the air and water. For instance, nano-based multifunctional electrical, optical, and chemical sensors can identify minute quantities of dangerous compounds. Nano-based adsorbents can, therefore, help remove harmful contaminants from the environment. Depending on the method used for surface functionalization, NPs can be used to create a wide range of unique smart materials that can aid in addressing the aforementioned global challenges. Nanomaterials' surface modification not only enhances their mechanical, chemical, and thermal characteristics but also gives them new and distinctive characteristics (B. Cashin et al., 2018; Khalid et al., 2020; Li et al., 2006; Rajendran et al., 2020; Simeonidis et al., 2016; Wang et al., 2020; Wieszczycka et al., 2021).

### 10.1 Preparation of Functionalized Polymers and Nanomaterials

Nanomaterials are synthesized using one of two methods. One of the important strategies is top-down methods, which incorporate technologies including mechanical milling, electrospinning, lithography, sputtering, arc discharge, and laser ablation. The second method uses bottom-up techniques such chemical vapor deposition (CVD), hydrothermal, solvothermal, and sol-gel processes as well as cationic, anionic, and polycondensation, *in situ*, and free radical polymerization.

DOI: 10.1201/9781003391364-10

Nanomaterials have demonstrated a range of unique characteristics that set them apart from their bulk counterparts. Large surface areas, quantum effects, antimicrobial activity, and superior thermal and electrical conductivities are among the characteristics of nanomaterials. The overall performance of metal-based catalysts is increased by improved dispersion of these catalysts on two-dimensional (2D) sheets and three-dimensional (3D) nanomaterials. Metal-based materials have shown extremely high catalytic activity at the nanoscale. The class of NPs includes carbon-based nanomaterials, nanoporous materials, core–shell materials, ultrathin 2D nanomaterials, and metal-based nanomaterials. NPs have an inorganic base, such as titanium dioxide ($TiO_2$), $SiO_2$, zinc oxide (ZnO), $Al_2O_3$, $Fe_2O_3$, and $Fe_3O_4$. Fullerenes, carbon nanotubes (CNTs), carbon-based quantum dots, graphene, and carbon nanohorns are examples of the fascinating class of carbon-based nanomaterials. Additionally, the surfaces of inorganic and carbon-based nanomaterials can be further functionalized to modify their properties for particular uses. Members of the family of carbon-based nanomaterials known as CNTs and graphene have attracted significant attention due to their large surface areas, rapid charge transfer abilities, and high mechanical strength. Despite the fact that ultrathin 2D materials are still being experimentally tested, they have remarkable characteristics and are being rapidly researched for useful applications. The fabrication of several metal-based nanostructured materials as nano-catalysts has made the synthesis of nanoscale catalysts a popular research area. The surface area, binding sites, and surface textural feature set of nanoscale catalysts are all quite large. The thermodynamics and kinetics of transportation during heterogeneous processes are benefited by all of these features and the tiny size. Investigated as electrode materials for energy conversion and storage devices are layered metal–oxide complexes. The efficacy of semiconductor metal–oxide materials in catalyzing water to produce sustainable energy is being improved by nanotuning. More attention is being placed on producing nanomaterials with controlled morphologies and nanoscale dimensions to achieve the desired outcomes due to well-organized nanostructures. Nanotechnology has already been used to introduce certain commercial products. In order to fulfil the increased energy demands of the future, nanomaterials will be integrated into next-generation gadgets. They will also play a more significant part in biosensors and nanomedicine to fight both current and emerging diseases. As an illustration, submicron $TiO_2$ particles are crucial to white paints. Commercial sunscreens comprise ZnO and $TiO_2$ NPs. Aside from that, it will take a lot of work to get most nanomaterials onto the commercial market because they are primarily being developed for lab-scale applications. The development of substitutes for the usage of endangered and limited-resource materials in nanomaterial production is a significant challenge associated with contemporary nanotechnology. There will be supply restrictions for 44 of 118 elements in the upcoming years. Precious metals, phosphorus, and rare-earth elements are examples of essential elements. It is necessary to reduce reliance on vital and endangered resources. Core–shell morphologies can be helpful in a range of applications to reduce the utilization of necessary components. Nanotechnology has the ability to contribute to wastewater recycling and water purification actively. A deeper understanding and the quick development of nanotechnology can be used to meet the challenges that modern society will confront in the future (Baig et al., 2021; Meng et al., 2016).

---

## 10.2 Functionalized Nanomaterials and Polymers in Wastewater Treatment

Membrane technology has steadily advanced in the wastewater treatment industry owing to its low capital cost, small equipment footprint, low power consumption, and excellent pollutant removal effectiveness. Traditional membranes have some drawbacks; however, functionalized nanomaterial membranes have proven to be a great alternative. For achieving higher effluent rejections, membranes with a hydrophilic nature, high water flux, fouling resistance, and a rough surface are essential. The efficiency of membrane filtering operations is significantly impacted by the surface modification technique. Nanomaterials are typically integrated into the membrane's bulk/support layer and/or surface/active layer. Numerous techniques (such as phase inversion, electrospinning, layer by layer, self-assembly method, interfacial polymerization, and etching) have been used to create membranes modified with nanomaterials that have enhanced characteristics. NPs were included into the polymeric matrix to produce the necessary surface textures, antifouling properties, and catalytic degradation. Heavy metals or organic effluents can be removed selectively using NP surface functional groups. Additionally, additional properties like antimicrobial activity, desalination, dairy demineralization, and dialysis separation can be acquired from the membrane with the right nanomaterial modification.

Despite the tremendous improvement, there are still numerous challenges to overcome.

(i)   At the moment, achieving homogeneous nanomaterial dispersion within membranes on a large scale is difficult. By choosing one or more effective combinations of functionalization techniques, creating large scales of high-quality, durable membranes should be possible.

(ii)  The hydrophobicity of the membrane surface can nevertheless be greatly reduced, especially for very harmful aqueous dispersants. Hydrophilicity can be significantly increased by immobilizing with zwitterions or creating a hydration layer by exposing $-NH_2$, $-COO$, and other groups on the membrane surface.

(iii) It is better to develop a reusable membrane that can filter out various forms of effluents and last for a long time without needing to be replaced. The performance of antifouling can be improved by modifying membranes with Ag, $TiO_2$ NPs, and other substances with intrinsic biocidal activity. This undoubtedly improved sustainability and antimicrobial resistance, but the toxicity of released metal ions toward aquatic animals is a serious concern.

(iv) A crucial component of wastewater purification is the detection and eradi-
cation of water-borne pathogens, such as viruses and bacteria. However,
there are not many articles on membrane technology that are focused on
eliminating microbes. Finally, it is possible to use more than one driving
force simultaneously to remove industrial effluents by combining membrane technology with other traditional processes, like precipitation and
coagulation. While heavy metals or other effluents can be retained in
the membrane, harmful dissolved compounds, for instance, can be precipitated. Future developments should be anticipated in order to achieve
improved selectivity and efficiency for wastewater treatment. As a result,
we believe our firm belief in membranes modified with nanomaterials as
exciting modalities for the removal of various dangerous effluents considerably aids in the ongoing development of clean water projects. It is
important to underline that although the main focus of this work is on the
use of nanomaterials in membrane technology for the efficient removal of
pollutants, advancements in the production of novel nanomaterials can also
be helpful in the development of membrane-based sensors and biosystems
(Díez & Rosal, 2020; Fonseca Couto et al., 2018; Khanzada et al., 2020;
Manikandan et al., 2022; Moradihamedani, 2022; Nain et al., 2022; Obotey
Ezugbe & Rathilal, 2020).

## 10.3 Functionalized Nanomaterials and Polymers in Desalination

The numerous and frequently one-of-a-kind characteristics of surface-modified/
functionalized nanocomposite membranes offer membrane technologists a wider
range of tools to custom-fabricate nanocomposite membranes with characteristics
that are best suited for a specific operation. Membranes made from surface-modified/
functionalized nanocomposites have the potential to be tailored to address all of the
problems associated with desalination and wastewater treatment, including fouling
and biofouling, while also extending the membrane's lifetime by enhancing mechanical robustness and resistance to cleaning regimens. None of this has an impact on
selectivity. We have made significant advancements in various aspects of life due to
our enhanced capacity for the monitoring, regulation, and manufacture of materials
at the nanoscale; these developments are now affecting the effectiveness and quality
of membrane materials used in water treatment. Without a doubt, the dissemination
of nanotechnological expertise will continue and positively affect the engineering
of membrane processes. To make these extremely reliable nanocomposite materials
the next generation of membranes, it is imperative to improve the interfacial interactions between organic polymers and surface-modified/functionalized nanocomposite membranes in the short term. As with other nanomaterials, further research
is needed to evaluate the environmental risks of use and to monitor the long-term
stability of these nanocomposites in practical operations. Before surface-modified/
functionalized membrane nanocomposites can be commercialized, more research is

required to make sure the advantages outweigh the fabrication and environmental costs (Al Aani et al., 2017; Jhaveri & Murthy, 2016; Lalia et al., 2013; Mohammad et al., 2015; Ng et al., 2013; Xu et al., 2015).

i. In order to address the growing need for freshwater worldwide, an emerging field has developed a nanomaterial for water treatment and desalination. Due to their natural adsorption and sieving abilities, which aid in removing minerals or contaminants from water, inorganic and carbon-based nanomaterials have attracted a lot of attention for use in the fabrication of membranes.

ii. A comprehensive review of the current state of the art in the fabrication of TiO₂, CNTs, and graphene oxide (GO)-embedded membranes revealed that novel approaches to achieving the highest level of contaminant removal effectiveness are emerging in a number of ways, including chemical and physical modifications of inorganic and carbon-based nanomaterials with other chemicals, such as sulfonic acids and aromatic groups.

iii. Carbon-based membranes have shown tremendous potential for desalination, and applications of inorganic and carbon-based nanomaterial-integrated membranes for the removal of inorganic impurities, emerging organic contaminants, and microbiological contaminants in water have been thoroughly explored with demonstrable efficiency.

iv. Greater antifouling properties are the result of the toxic effects of inorganic and carbon-based nanomaterial-integrated membranes, which destroy microorganisms by producing reactive oxygen species, and the high electrical conductivity, hydrophilicity, and positive/negative charges on the surface of nanomaterials, which allow these membranes to repel foulants.

v. Despite the significant effort made to understand the function, structure, properties, and synthesis of inorganic and carbon-based nanomaterials, many problems still need to be resolved. In addition, there is still considerable potential for improvement in the use of inorganic and carbon-based nanomaterial-integrated membranes for desalination and water treatment (Ali et al., 2019).

## 10.4 Functionalized Nanomaterials and Polymers in Bioremediation

In particular, nanocomposites and hybrid materials based on the three distinct kinds of nanomaterials—metal oxide NPs, carbon-based nanomaterials, and metal NPs—are highlighted as recent prospective solutions for environmental treatment methods. The various nanomaterials used as nanofillers provide the composite materials' catalytic (or photocatalytic), redox, sorption/desorption, and magnetic properties. There are numerous instances where different NPs have been combined to produce functional composite materials with all the characteristics of the initial nanofillers and entirely novel properties. One advantage of the composite materials that have been shown is their ability to regenerate and reuse with little efficiency loss even after several adsorption cycles. It also explains how they can destroy organic pollutants by catalytic, photocatalytic, or electrochemical processes.

Using bio-based polymers and useful nanomaterials, environment-friendly synthesis methods and strategies for developing more durable materials have been established. The *in situ* and *ex situ* bioremediation approaches, including microbial fuel cells, and the immobilization of bacteria or enzymes for deteriorating processes are most notable for highlighting the potential applications of these novel materials and solutions (Fu & Wang, 2011; Ihsanullah et al., 2016; Rando et al., 2022).

## 10.5 Functionalized Nanomaterials and Polymers in Energy Storage

In polymer electrolyte membrane (PEM) research for direct methanol fuel cells (DMFCs) and proton exchange membrane fuel cells (PEMFCs), CNTs are frequently used. Particularly at high temperatures and low humidity, functionalized polymers, inorganic, and carbon-based nanomaterials with a variety of functional groups, such as acidic groups (sulfonic acid, phosphonic acid, and carboxylic acid) and basic groups (e.g., amino ($-NH_2$)), could increase proton conductivity via the Grotthuss mechanism. The compatibility and dispersion of composite membranes including CNT fillers coated with $SiO_2$, $TiO_2$, and SZr are very high, and the coatings could hinder van der Waals interactions of CNTs and reduce their surface tension. The durability, water retention, methanol permeability, and thermal stability of fuel cells are all enhanced by incorporating CNTs into most polymer matrices. Comparing the performance of non-fluoropolymer-based composite membranes to Nafion-based composite membranes under high temperature and low relative humidity conditions reveals higher performance advantages, and interest in these materials has grown due to their low cost and non-polluting characteristics. Contrarily, composite membranes containing inorganic and carbon-based nanomaterials may cause an insignificant decrease in proton conductivity, especially for composite membranes not based on fluoropolymers that exhibit poor proton conductivity findings. This review shows that functionalized nanomaterials significantly increase the membranes' mechanical and thermal stability, but the addition of nanomaterial fillers reduces the membranes' proton conductivity. As a result, functionalized inorganic and carbon-based nanomaterial composite membrane fuel cells still lack the commercial viability required to achieve these results (Esmaeili et al., 2019; Fan et al., 2021; Gao et al., 2023; Okonkwo et al., 2021; Vinothkannan et al., 2021). As a ground-breaking photovoltaic technology, dye-sensitized solar cells (DSSCs) have the potential to compete with traditional solar cells. $TiO_2$, one of the materials used in DSSCs, is often affordable, abundant, and eco-friendly. They can move faster from the research lab to the mass production line because they are less sensitive to manufacturing flaws than silicon solar cells. For applications, portable electronic devices prefer the light weight and flexibility of DSSCs. It has been found that DSSCs outperform silicon solar cells in low-light situations, like at dawn and twilight, and that high temperatures have no negative effects on their overall efficiency. Additionally, the transparency and changeable color of DSSCs could be applied to ornamental elements like windows and sunroofs. Such advantages have so far drawn a lot of money from both the government and corporate fundings. With continued work, it is anticipated that DSSCs will serve as a dependable source of electricity in the future (Gong et al., 2012; Grätzel, 2003).

## 10.6  Functional Nanomaterials: Industrial Potential and Challenges

Nanotechnology is an industrially relevant enabling technology that, via the integration of science and technology, is predicted to produce major new product/market potential. This will revitalize the chemical industry, and improvements in nanotechnology will also benefit most of today's industries. Globally broad partnerships, collaborations, and networks between industrial, academic, and governmental organizations will promote the development of new products. Traditional, large-volume manufacturing-based business models are obviously not always applicable because the majority of value will be created as specialist, low-volume products, with licensable patents making up a sizable portion of the total value created. It is anticipated that cleaner, more energy-efficient processes; new renewable resources as feedstocks; and less environmental impact from industrial facilities will all benefit society (Qiu Zhao et al., 2003).

Molecular NPs are used in a various industrial processes. They are crucial for a variety of industrial applications because of their size. In the industrial setting, NPs are employed to enhance the client-supplied product. Food-based companies improved food quality with exceptional flavor and aroma by using NPs. In agriculture, NPs are employed as insecticides to make fields productive and to lessen their toxicity. The textile industry may develop specialized fabrics with a range of uses using NPs. The toxicity of cosmetic items can be reduced by using NPs. Other industries, including those focused on energy, water, and the environment as well as oil and gas, are now using NPs to create products and solve issues. Since they can help minimize the usage of conventional materials by improving the presentation of development materials and lowering/reducing the use of energy, NPs are utilized in industrial development. Nanotechnology is now used in many aspects of modern life, particularly in the fields of electronics, space, healthcare, food, cosmetics, composites, and energy. As a result, it has emerged as a crucially important innovation for a wide range of applications and has become a major concern for the advancement of science and innovation strategies. With the exception of energy, where the results have been remarkable, nanotechnology has so far had a considerable impact on nanomaterial synthesis. However, research work must be done to expand its penetration at the system level. In conclusion, molecular nanomaterials are revolutionizing a variety of industries, including the textile, gas and oil, food, and computer industry. The long-term growth of novel nanomaterial-based industrial applications and consumer benefits across the nation implicate their impact on human life, the environment, and the overall process of modern civilization, an internationally recognized, rigid, nanosafety regulation (Dąbrowska et al., 2018; Hosne Asif & Hasan, 2018; Subhan et al., 2021).

## 10.7 Challenges of Functionalized Nanomaterials in the Environment

The main obstacles of smart material applications, however, are associated with converting these concepts or prototypes into final products. The industrial production of smart nanomaterials presents a number of challenges, including scaling up, the overall response to external variable conditions, the degradation and toxicity of nanoproducts, and the high cost of testing before final production, which includes approvals and documentation, legalization, administrative costs, stability studies, and other factors. The prolonged contact of smart NPs with the environment and humans is another issue. Despite the many benefits of smart nanomaterials, some people are concerned about their toxicity, the possibility that humans and animals might inhale them in, and the possibility that aquatic life could absorb them from water. More research is needed to solve these obstacles in the future (Wieszczycka et al., 2021).

## References

Al Aani, S., Wright, C. J., Atieh, M. A., & Hilal, N. (2017). Engineering nanocomposite membranes: Addressing current challenges and future opportunities. *Desalination, 401*, 1–15. https://doi.org/10.1016/j.desal.2016.08.001

Ali, S., Rehman, S. A. U., Luan, H.-Y., Farid, M. U., & Huang, H. (2019). Challenges and opportunities in functional carbon nanotubes for membrane-based water treatment and desalination. *Science of the Total Environment, 646*, 1126–1139. https://doi.org/10.1016/j.scitotenv.2018.07.348

Baig, N., Kammakakam, I., & Falath, W. (2021). Nanomaterials: A review of synthesis methods, properties, recent progress, and challenges. *Materials Advances, 2*(6), 1821–1871. https://doi.org/10.1039/D0MA00807A

Cashin, V. B., Eldridge, D. S., Yu, A., & Zhao, D. (2018). Surface functionalization and manipulation of mesoporous silica adsorbents for improved removal of pollutants: A review. *Environmental Science: Water Research & Technology, 4*(2), 110–128. https://doi.org/10.1039/C7EW00322F

Dąbrowska, S., Chudoba, T., Wojnarowicz, J., & Łojkowski, W. (2018). Current trends in the development of microwave reactors for the synthesis of nanomaterials in laboratories and industries: A review. *Crystals, 8*(10), Article 10. https://doi.org/10.3390/cryst8100379

Díez, B., & Rosal, R. (2020). A critical review of membrane modification techniques for fouling and biofouling control in pressure-driven membrane processes. *Nanotechnology for Environmental Engineering, 5*(2), 15. https://doi.org/10.1007/s41204-020-00077-x

Esmaeili, N., Gray, E. MacA., & Webb, C. J. (2019). Non-fluorinated polymer composite proton exchange membranes for fuel cell applications—A review. *ChemPhysChem, 20*(16), 2016–2053. https://doi.org/10.1002/cphc.201900191

Fan, L., Tu, Z., & Chan, S. H. (2021). Recent development of hydrogen and fuel cell technologies: A review. *Energy Reports, 7*, 8421–8446. https://doi.org/10.1016/j.egyr.2021.08.003

Fonseca Couto, C., Lange, L. C., & Santos Amaral, M. C. (2018). A critical review on membrane separation processes applied to remove pharmaceutically active compounds from water and wastewater. *Journal of Water Process Engineering, 26*, 156–175. https://doi.org/10.1016/j.jwpe.2018.10.010

Fu, F., & Wang, Q. (2011). Removal of heavy metal ions from wastewaters: A review. *Journal of Environmental Management, 92*(3), 407–418. https://doi.org/10.1016/j.jenvman.2010.11.011

Gao, J., Dong, X., Tian, Q., & He, Y. (2023). Carbon nanotubes reinforced proton exchange membranes in fuel cells: An overview. *International Journal of Hydrogen Energy, 48*(8), 3216–3231. https://doi.org/10.1016/j.ijhydene.2022.10.173

Gong, J., Liang, J., & Sumathy, K. (2012). Review on dye-sensitized solar cells (DSSCs): Fundamental concepts and novel materials. *Renewable and Sustainable Energy Reviews, 16*(8), 5848–5860. https://doi.org/10.1016/j.rser.2012.04.044

Grätzel, M. (2003). Dye-sensitized solar cells. *Journal of Photochemistry and Photobiology C: Photochemistry Reviews, 4*(2), 145–153. https://doi.org/10.1016/S1389-5567(03)00026-1

Hosne Asif, A. K. M. A., & Hasan, Md. Z. (2018). Application of nanotechnology in modern textiles: A review. *International Journal of Current Engineering and Technology, 8*(2). https://doi.org/10.14741/ijcet/v.8.2.5

Ihsanullah, Abbas, A., Al-Amer, A. M., Laoui, T., Al-Marri, M. J., Nasser, M. S., Khraisheh, M., & Atieh, M. A. (2016). Heavy metal removal from aqueous solution by advanced carbon nanotubes: Critical review of adsorption applications. *Separation and Purification Technology, 157*, 141–161. https://doi.org/10.1016/j.seppur.2015.11.039

Jhaveri, J. H., & Murthy, Z. V. P. (2016). A comprehensive review on anti-fouling nanocomposite membranes for pressure driven membrane separation processes. *Desalination, 379*, 137–154. https://doi.org/10.1016/j.desal.2015.11.009

Khalid, K., Tan, X., Mohd Zaid, H. F., Tao, Y., Lye Chew, C., Chu, D.-T., Lam, M. K., Ho, Y.-C., Lim, J. W., & Chin Wei, L. (2020). Advanced in developmental organic and inorganic nanomaterial: A review. *Bioengineered, 11*(1), 328–355. https://doi.org/10.1080/21655979.2020.1736240

Khanzada, N. K., Farid, M. U., Kharraz, J. A., Choi, J., Tang, C. Y., Nghiem, L. D., Jang, A., & An, A. K. (2020). Removal of organic micropollutants using advanced membrane-based water and wastewater treatment: A review. *Journal of Membrane Science, 598*, 117672. https://doi.org/10.1016/j.memsci.2019.117672

Lalia, B. S., Kochkodan, V., Hashaikeh, R., & Hilal, N. (2013). A review on membrane fabrication: Structure, properties and performance relationship. *Desalination, 326*, 77–95. https://doi.org/10.1016/j.desal.2013.06.016

Li, L., Fan, M., Brown, R. C., Van Leeuwen, J. (Hans), Wang, J., Wang, W., Song, Y., & Zhang, P. (2006). Synthesis, properties, and environmental applications of nanoscale iron-based materials: A review. *Critical Reviews in Environmental Science and Technology, 36*(5), 405–431. https://doi.org/10.1080/10643380600620387

Manikandan, S., Subbaiya, R., Saravanan, M., Ponraj, M., Selvam, M., & Pugazhendhi, A. (2022). A critical review of advanced nanotechnology and hybrid membrane based water recycling, reuse, and wastewater treatment processes. *Chemosphere, 289*, 132867. https://doi.org/10.1016/j.chemosphere.2021.132867

Meng, L.-Y., Wang, B., Ma, M.-G., & Lin, K.-L. (2016). The progress of microwave-assisted hydrothermal method in the synthesis of functional nanomaterials. *Materials Today Chemistry, 1–2*, 63–83. https://doi.org/10.1016/j.mtchem.2016.11.003

Mohammad, A. W., Teow, Y. H., Ang, W. L., Chung, Y. T., Oatley-Radcliffe, D. L., & Hilal, N. (2015). Nanofiltration membranes review: Recent advances and future prospects. *Desalination, 356*, 226–254. https://doi.org/10.1016/j.desal.2014.10.043

Moradihamedani, P. (2022). Recent advances in dye removal from wastewater by membrane technology: A review. *Polymer Bulletin, 79*(4), 2603–2631. https://doi.org/10.1007/s00289-021-03603-2

Nain, A., Sangili, A., Hu, S.-R., Chen, C.-H., Chen, Y.-L., & Chang, H.-T. (2022). Recent progress in nanomaterial-functionalized membranes for removal of pollutants. *IScience, 25*(7), 104616. https://doi.org/10.1016/j.isci.2022.104616

Ng, L. Y., Mohammad, A. W., Leo, C. P., & Hilal, N. (2013). Polymeric membranes incorporated with metal/metal oxide nanoparticles: A comprehensive review. *Desalination, 308*, 15–33. https://doi.org/10.1016/j.desal.2010.11.033

Obotey Ezugbe, E., & Rathilal, S. (2020). Membrane technologies in wastewater treatment: A review. *Membranes, 10*(5), Article 5. https://doi.org/10.3390/membranes10050089

Okonkwo, P. C., Ben Belgacem, I., Emori, W., & Uzoma, P. C. (2021). Nafion degradation mechanisms in proton exchange membrane fuel cell (PEMFC) system: A review. *International Journal of Hydrogen Energy, 46*(55), 27956–27973. https://doi.org/10.1016/j.ijhydene.2021.06.032

Qiu Zhao, Q., Boxman, A., & Chowdhry, U. (2003). Nanotechnology in the chemical industry—Opportunities and challenges. *Journal of Nanoparticle Research, 5*(5), 567–572. https://doi.org/10.1023/B:NANO.0000006151.03088.cb

Rajendran, A., Rajendiran, M., Yang, Z.-F., Fan, H.-X., Cui, T.-Y., Zhang, Y.-G., & Li, W.-Y. (2020). Functionalized silicas for metal-free and metal-based catalytic applications: A review in perspective of green chemistry. *The Chemical Record, 20*(6), 513–540. https://doi.org/10.1002/tcr.201900056

Rando, G., Sfameni, S., Galletta, M., Drommi, D., Cappello, S., & Plutino, M. R. (2022). Functional nanohybrids and nanocomposites development for the removal of environmental pollutants and bioremediation. *Molecules, 27*(15), Article 15. https://doi.org/10.3390/molecules27154856

Simeonidis, K., Mourdikoudis, S., Kaprara, E., Mitrakas, M., & Polavarapu, L. (2016). Inorganic engineered nanoparticles in drinking water treatment: A critical review. *Environmental Science: Water Research & Technology, 2*(1), 43–70. https://doi.org/10.1039/C5EW00152H

Subhan, M. A., Choudhury, K. P., & Neogi, N. (2021). Advances with molecular nanomaterials in industrial manufacturing applications. *Nanomanufacturing, 1*(2), Article 2. https://doi.org/10.3390/nanomanufacturing1020008

Vinothkannan, M., Rhan Kim, A., & Jin Yoo, D. (2021). Potential carbon nanomaterials as additives for state-of-the-art Nafion electrolyte in proton-exchange membrane fuel cells: A concise review. *RSC Advances, 11*(30), 18351–18370. https://doi.org/10.1039/D1RA00685A

Wang, H., Liang, X., Wang, J., Jiao, S., & Xue, D. (2020). Multifunctional inorganic nano-materials for energy applications. *Nanoscale, 12*(1), 14–42. https://doi.org/10.1039/C9NR07008G

Wieszczycka, K., Staszak, K., Woźniak-Budych, M. J., Litowczenko, J., Maciejewska, B. M., & Jurga, S. (2021). Surface functionalization—The way for advanced applications of smart materials. *Coordination Chemistry Reviews, 436*, 213846. https://doi.org/10.1016/j.ccr.2021.213846

Xu, G.-R., Wang, S.-H., Zhao, H.-L., Wu, S.-B., Xu, J.-M., Li, L., & Liu, X.-Y. (2015). Layer-by-layer (LBL) assembly technology as promising strategy for tailoring pressure-driven desalination membranes. *Journal of Membrane Science, 493*, 428–443. https://doi.org/10.1016/j.memsci.2015.06.038

# Index

For Product Safety Concerns and Information please contact our EU
representative  GPSR@taylorandfrancis.com
Taylor & Francis Verlag GmbH, Kaufingerstraße 24, 80331 München, Germany

www.ingramcontent.com/pod-product-compliance
Lightning Source LLC
Chambersburg PA
CBHW070726220326
41598CB00024BA/3324